The scientific and technological importance of lasers has generated great interest in the area of cavity nonlinear optics. This book provides a thorough description of the field in terms of modern dynamical systems theory. Throughout, the emphasis is on deriving analytical results and highlighting their physical significance.

The early chapters introduce the physical models for active and passive cavities. In later chapters, these models are applied to a variety of problems in laser theory, optical bistability, and parametric oscillators. Subjects covered include scaling laws, Hopf bifurcations, passive Q-switching, and Turing instabilities. Several of the topics treated cannot be found in other books, such as swept control parameter dynamics, laser stability, multimode rate equations, and antiphase dynamics.

The book stresses the connections between theoretical work and actual experimental results and will be of great interest to graduate students and researchers in theoretical physics, nonlinear optics, and laser physics.

T0275560

CAMBRIDGE STUDIES IN MODERN OPTICS

Series Editors

P. L. KNIGHT

Department of Physics,
Imperial College of Science, Technology, and Medicine

A. MILLER

Department of Physics and Astronomy, University of St. Andrews

Theoretical Problems in Cavity Nonlinear Optics

TITLES IN PRINT IN THIS SERIES

Theoretical Problems in Cavity Nonlinear Optics

PAUL MANDEL

Fonds National de la Recherche Scientifique and Université Libre de Bruxelles

CAMBRIDGE UNIVERSITY PRESS

CAMBRIDGE UNIVERSITY PRESS
Cambridge, New York, Melbourne, Madrid, Cape Town, Singapore, São Paulo

Cambridge University Press
The Edinburgh Building, Cambridge CB2 2RU, UK

Published in the United States of America by Cambridge University Press, New York

www.cambridge.org
Information on this title: www.cambridge.org/9780521553858

© Cambridge University Press 1997

First published 1997
This digitally printed first paperback version 2005

A catalogue record for this publication is available from the British Library

Library of Congress Cataloguing in Publication data

Mandel, P. (Paul), 1942–
Theoretical problems in cavity nonlinear optics / Paul Mandel.
p.   cm. – (Cambridge studies in modern optics)
Includes index.
ISBN 0-521-55385-7
1. Nonlinear optics.   2. Optical bistability.   3. Lasers.
I. Title.   II Series.
QC446.2.M37   1997   96-15181
535′.2–dc20   CIP

ISBN-13 978-0-521-55385-8 hardback
ISBN-10 0-521-55385-7 hardback

ISBN-13 978-0-521-01920-0 paperback
ISBN-10 0-521-01920-6 paperback

To all those who gave their life
for the freedom to think, to write, and to read.

Contents

Introduction

Nonlinear optics is a fairly young science, having taken off with the advent of the laser in 1960. Nonlinear optics (NLO) deals with the interaction of electromagnetic waves and matter in the infrared, visible, and ultraviolet domains. The frontiers of NLO are somewhat blurred, but microwaves and γ-rays are clearly outside its domain. This book is entirely devoted to a study of NLO in a resonant cavity and when the quantum nature of the electromagnetic field is not of prime importance. More precisely, we study those aspects of cavity NLO in which fluctuations in the number of photons and atoms are not relevant. This particular area of optics is dominated by the Maxwell–Bloch equations, which constitute its paradigm in the sense of T. S. Kuhn.[1] The status of the Maxwell–Bloch equations is quite peculiar. From a fundamental viewpoint, they describe the laws of evolution of the first moments of a density operator, which verifies the von Neumann equation. However, to account for the finite lifetime of the atoms and of the field in the necessarily lossy cavity, some legerdemains have to be introduced to obtain the Maxwell–Bloch equations. Stated more explicitly, the von Neumann equation for a large but finite system does not explain irreversibility, whereas the Maxwell–Bloch equations fully include the irreversible decay of the atoms and of the cavity field. This problem is not specifically related to optics but reflects the general failure of statistical mechanics to explain convincingly the irreversible evolution of macroscopic systems. What makes this difficulty more critical in cavity NLO is that the atomic decay rates are fundamental parameters: The problems discussed in this book would not exist at all if atoms and fields were stable.

There are two ways to consider the Maxwell–Bloch equations—both proving that it is reasonable to give them the paradigmatic status. One approach is a

---

[1] T. S. Kuhn, *The Structure of Scientific Revolutions,* 2nd ed. (The University of Chicago Press, Chicago, 1970).

derivation from first principles (the von Neumann equation) and an attempt to justify the introduction of dissipation, that is, irreversibility, through incredibly elaborate theoretical developments. Hundreds of papers and many books have been devoted to this justification. Unfortunately, this approach suffers from the same weakness as the derivations of irreversible equations in statistical mechanics. The other attitude is to accept this weakness of the Maxwell–Bloch equations and to convince the scientific community that these equations have value in explaining and predicting phenomena. Here again, two options are possible. The general trend is to apply the Maxwell–Bloch equations to as broad a class of problems as possible. This often implies a lack of depth in the mathematical analysis in exchange for the large spectrum of problems that are shown to fit nicely into the realm of the paradigm. The other alternative is to look at a small number of problems but analyze them in depth. To quote T. S. Kuhn, "So long as the tools a paradigm supplies continue to prove capable of solving the problems it defines, science moves fastest and penetrates most deeply through confident employment of those tools."[2] The purpose of this book is to show that the Maxwell–Bloch equations do indeed furnish powerful tools that can be used to describe with amazing detail many phenomena occurring in cavity NLO. This explains why the derivation of the Maxwell–Bloch equations is only sketched, whereas the emphasis of the book is on a systematic analysis of selected cavity NLO problems. The reader is assumed to have had prior exposure to laser physics. Textbooks of which this book tries to be an extension are the following:

- H. Haken. Laser Theory in *Encyclopedia of Physics*. L. Genzel, ed. (Springer, Heidelberg, 1970)
- M. Sargent III, M. O. Scully, and W. Lamb, Jr. *Laser Physics* (Addison-Wesley, Reading, 1974)
- H. Haken. *Light,* Vols. 1 & 2 (North-Holland, Amsterdam, 1981).
- Ya. I. Khanin. *Principles of Laser Dynamics* (Elsevier, Amsterdam, 1995).

Many more textbooks are available in NLO. However, they reflect other priorities than those that have been my guide in the preparation of this text. Although this book does not require a vast culture in mathematics, it surely will help if the reader has discovered that mathematics is the language of Nature.

This book is structured as follows. The first chapter sets the décors and the priorities for the remainder of the book. Chapters 2 and 3 deal with the problem of swept parameter across the laser first threshold. In Chapter 2, only the simplest situation is considered. A systematic simplification of the laser model

---

[2] T. S. Kuhn, op. cit. p. 76.

is performed to arrive at the conclusion that a local 1-D model captures the essential aspects of the problem. In Chapter 3, I extend this analysis by considering the influence of the initial condition and of an external field of constant, periodic, or stochastic amplitude.

Chapters 4 to 6 are devoted to optical bistability. In Chapter 4, I derive the basic equations for single-mode optical bistability and I show again that a reduction to generic 1-D models is possible in a number of situations that are relevant for the experiments. I also consider the response of a bistable device to input pulses. In Chapter 5, I consider the response of a bistable system near its limit point either to a ramp or to a periodic modulation of the input field. Finally I introduce a model for nascent hysteresis that serves as a basis for the optical transistor theory. Chapter 6 is entirely devoted to multimode optical bistability à la Ikeda.

Chapters 7 and 8 deal with the topic of rate equations for free-running multimode lasers. In Chapter 7, I discuss the derivation of the rate equations for the ring and the Fabry–Pérot lasers. Chapter 8 is an introduction to antiphase dynamics. The remaining four chapters are the mathematical chapters. They are more difficult to read, and the material presented in these chapters is usually not covered in other books on NLO.

Chapter 9 deals with the important question of the laser stability when the ouput intensity is constant in time. Although the approach followed in this chapter (and, for that matter, in this whole book) does not pertain to the science of the engineer, an insight into the laser stability can be obtained from the consideration of models that can be solved exactly. These models cover a surprisingly wide class of lasers.

Chapter 10 is an introduction to the Hopf bifurcation analysis. This topic is never treated at the right level for most physicists. I have tried to remedy this situation by an iterative presentation, showing step by step how problems occur and how they can be fixed. It turns out that second harmonic generation offers the simplest example of a Hopf bifurcation in cavity NLO.

Chapter 11 introduces the reader to composite cavities and their rich physics. It also serves a double mathematical purpose. It provides another example on which the technique of the Hopf bifurcation developed in Chapter 10 can be applied. It also provides an example of passive Q-switching that can be solved completely by the method of matched asymptotic expansions.

Chapter 12 deals with transverse effects. The emphasis has been placed on two aspects of this problem: how to derive from the Maxwell–Bloch equations a slowly varying amplitude equation (Ginzburg–Landau or Swift–Hohenberg, for instance) and how to analyze a Turing bifurcation from homogeneous to inhomogeneous solutions.

I fully realize that the consideration of a few problems, selected because they lend themselves to an analytic treatment, does not necessarily help the reader to solve his or her particular problem. However, it is my hope that model calculations serve a purpose by showing how to handle some of the standard problems that arise in NLO. The problems solved in the last four chapters were still considered to be insoluble fifteen years ago. Those related to swept parameter were not even acknowledged as problems ten years ago in the optical community.

I am indebted to Prof. J. Y. Zhang, who invited me to present in the fall of 1993 a set of lectures on cavity NLO at Northwest University, in Xi'an (China). This book has its roots in the lecture notes I prepared for this course and in stimulating discussions with T. Erneux. I also wish to thank Prof. M. Lallemant, who gave me a second chance to present the content of this book in the form of lectures given in the fall of 1995 at the Université de Bourgogne in Dijon (France). Finally, it is a pleasure to acknowledge G. Lythe, B. A. Nguyen, D. Pieroux, M. Tlidi, and J. Y. Wang, who kindly checked parts of the manuscript. This introduction is a unique opportunity to express my gratitude to the Fonds National de la Recherche Scientifique – and its General Secretaries, P. Levaux and M. J. Simoen – which has been my employer throughout my scientific life and has provided me with the rare opportunity of a tenured and full-time research position. Research without grants is not an easy task these days. I have been spared the burden of spending much of my time looking for grants thanks to three major grants, one from the European Commission through its Stimulation Action Framework Program managed by the DG XII (1984–86) and two consecutive grants from the Inter-University Attraction Pole Program of the Belgian government (1987–1996). Finally, my alma mater deserves special thanks for accepting to keep under its roof a singularity that is not always easy to live with.

P. Mandel
Brussels, December 1995

1

Reduction of the Maxwell–Schrödinger equations

A rigorous description of the light–matter interaction requires the use of quantized Maxwell equations to describe the light field and the Schrödinger equation to describe the material medium. This book does not attempt to deal with quantum optical problems that necessitate a field quantization. Instead we focus on properties that are generally grouped under the umbrella of *nonlinear optics*. Although many definitions of this expression exist, we use it in the sense that the properties we deal with do not depend in a crucial way on the field quantization. For practical purposes, this means in general that the average photon number is large and that we can neglect spontaneous emission as a dynamical process. Rather, we include the consequence of spontaneous emission, that is, the instability of the atomic levels, in a purely phenomenological way. When this is done, we call the material equations the Bloch equations instead of the Schrödinger equation. References [1] through [4] are the classical textbooks that contain a tentative justification for the transition from the Schrödinger to the Bloch equations. None of these attempts is satisfactory from a fundamental viewpoint because the real difficulty is to incorporate the finite lifetime (or natural linewidth) of the atomic energy levels in the Schrödinger equation. This remains an open problem as of now.

1.1 The Maxwell–Schrödinger equations

We consider atoms interacting with an intense electromagnetic field. By *intense* we mean a field for which the quantization is not necessary. Hence the system is described by Maxwell's equations for the classical electric field, E

$$c^2 \partial^2 E/\partial z^2 - \partial^2 E/\partial t^2 = \varepsilon_0^{-1} \partial^2 P/\partial t^2 \qquad (1.1)$$

In this equation, P is the macroscopic atomic polarization of the medium induced by the interaction with the field, $c = (\mu_0 \varepsilon_0)^{-1/2}$ is the speed of light, and μ_0 and ε_0 are the linear (that is, the field-independent) magnetic permeability and the dielectric (or permittivity) constants, respectively. We have already simplified the problem by reducing the space dependence to only one dimension and by neglecting the vectorial character of the field. Each atom is described by a Schrödinger equation

$$i\hbar\,\partial\Psi/\partial t = H\Psi, \qquad H = H_0 + V, \quad V = \alpha E \qquad (1.2)$$

where $\alpha = \alpha(e\mathbf{r})$ is the projection of the electric dipole moment in the direction of the field polarization. The atoms are described as two-level atoms. Hence, their wave function is

$$\Psi = A\psi_0 + B\psi_1, \quad |A|^2 + |B|^2 = 1, \quad H_0\psi_j = \hbar\omega_j\psi_j, \quad j = 0, 1$$

$$i\,\partial A/\partial t \cong \omega_0 A + (\mu E/\hbar)B + (\nu_0 E/\hbar)A \simeq \omega_0 A + (\mu E/\hbar)B$$

$$i\,\partial B/\partial t \cong \omega_1 B + (\mu E/\hbar)A + (\nu_1 E/\hbar)B \simeq \omega_1 B + (\mu E/\hbar)A \qquad (1.3)$$

In equations (1.3), μ is the matrix element of the electric dipole moment between the two states ψ_0 and ψ_1, whereas ν_j is the matrix element of the electric dipole moment between the same state ψ_j. The matrix element ν_j vanishes if the atomic medium is invariant under a reflexion symmetry, implying $|\psi_j(\mathbf{r})|^2 = |\psi_j(-\mathbf{r})|^2$. This is the general rule and it will be assumed to hold here. A notable exception are crystals lacking a center of symmetry (see Chapters 10 and 11). Other possible exceptions are spatially inhomogeneous systems such as phase boundaries and, in particular, solid–gas interfaces. The approximations made in deriving the Schrödinger amplitude equations, (1.3), for A and B imply that we consider only one-photon transitions and neglect the center-of-mass dynamics. The other assumption made in deriving equations (1.3) is to neglect the variation of the electric field within the atom. This introduces the dipole approximation

$$\int \psi_p^*(\mathbf{r})E(\mathbf{r}, t)\alpha(e\mathbf{r})\psi_q(\mathbf{r})\,d\mathbf{r} \cong E(\mathbf{r}_a, t)\int \psi_p^*(\mathbf{r})\alpha(e\mathbf{r})\psi_q(\mathbf{r})\,d\mathbf{r}$$

$$\equiv \mu E(\mathbf{r}_a, t), \qquad p \neq q \qquad (1.4)$$

where $\mathbf{r}_a$ is the position of the atomic center-of-mass around which the atomic wave function is sharply peaked. Without loss of generality for the following analysis, we may take μ as a real parameter.

The material medium is composed of many atoms, and what we need to feed into Maxwell's equation is the macroscopic atomic polarization. Here again, we introduce a drastic simplification. Namely, we assume that all atoms contribute in the same way to the polarization. This means that the center-of-mass motion and the dispersion of the atomic frequencies are neglected. These are reasonable assumptions for most solid-state lasers. However, in Chapter 9, which deals with the problem of laser stability, we restore the atomic frequency dispersion and determine its influence on the instability threshold. For gas lasers a more sophisticated statistical average is required to account for the center-of-mass velocity distribution.

We denote by angle brackets ensemble averages taken over all the atoms. As an example, let us compute two mean values that we will need later, using the Schrödinger amplitude equations, (1.3),

$$\partial\langle A^*B\rangle/\partial t = -i(\omega_1 - \omega_0)\langle A^*B\rangle + i(\mu/\hbar)[\langle E|B|^2\rangle - \langle E|A|^2\rangle] \quad (1.5)$$

$$\partial\langle|A|^2\rangle/\partial t = -\partial\langle|B|^2\rangle/\partial t = i(\mu/\hbar)\langle EAB^*\rangle - i(\mu/\hbar)\langle EA^*B\rangle \quad (1.6)$$

where * is the complex conjugation operation. At this point, we introduce the approximation that is probably the most difficult to justify. We assume that the statistical average of a function P of the field and a function Q of the atomic variables can be factored into the product of the two averages

$$\langle P_{\text{field}}Q_{\text{atom}}\rangle \cong \langle P_{\text{field}}\rangle\langle Q_{\text{atom}}\rangle \simeq P_{\text{field}}\langle Q_{\text{atom}}\rangle \quad (1.7)$$

The approximation $\langle P_{\text{field}}\rangle \simeq P_{\text{field}}$ means that the discrete character of the atomic center-of-mass distribution is neglected and functions of the field E are treated as functions of the continuous variable $\mathbf{r}$. A justification of the approximation (1.7) requires a full statistical treatment of the problem and will not be attempted here. The difficulty of this problem has not made it popular and the scientific literature does not abound with papers dealing with the justification of this assumption.

With this simplification, we have

$$\partial\langle A^*B\rangle/\partial t = -i\omega_a\langle A^*B\rangle + i(\mu/\hbar)E[\langle|B|^2\rangle - \langle|A|^2\rangle] \quad (1.8)$$

$$\partial\langle|A|^2\rangle/\partial t = i(\mu/\hbar)E[\langle AB^*\rangle - \langle A^*B\rangle] \quad (1.9)$$

where we have introduced the atomic frequency $\omega_a = \omega_1 - \omega_0$. From this result, we derive an equation for the atomic population difference $\mathcal{D} \equiv \langle|A|^2\rangle - \langle|B|^2\rangle$

$$\partial\mathcal{D}/\partial t = -(2i\mu/\hbar)E(\mathcal{P} - \mathcal{P}^*) \quad (1.10)$$

where $\mathcal{P} = \langle A^*B\rangle$. We choose to use the complex function $\mathcal{P}$ instead of the real atomic polarization $p \equiv \mu(\langle A^*B\rangle + \langle AB^*\rangle)$ because it leads to a simpler formulation of the dissipative material evolution equations. With the normalization $\langle|A|^2\rangle + \langle|B|^2\rangle = 1$, $\mathcal{D}$ is the population difference divided by the number of atoms, N, and p is the atomic polarization divided by N. The function $\mathcal{P}$ satisfies the equation

$$\partial \mathcal{P}/\partial t = -i\omega_a \mathcal{P} - (i\mu/\hbar)E\mathcal{D} \tag{1.11}$$

The complex polarization $\mathcal{P}$ that appears in the material equation, (1.10), for the population difference is induced by the electric field E acting upon the atoms. The self-consistency of the semiclassical theory of nonlinear optics rests on the requirement that the total atomic polarization $N\mu(\mathcal{P} + \mathcal{P}^*)$ induced by the electric field E is identical to the macroscopic polarization P that is the source of the electric field E in Maxwell's equation, (1.1). Whence we obtain a closed set of coupled light–matter equations

$$c^2 \partial^2 E/\partial z^2 - \partial^2 E/\partial t^2 = (N\mu/\varepsilon_0)\partial^2(\mathcal{P} + \mathcal{P}^*)/\partial t^2 \tag{1.12}$$

$$\partial \mathcal{P}/\partial t = -i\omega_a \mathcal{P} - (i\mu/\hbar)E\mathcal{D} \tag{1.13}$$

$$\partial \mathcal{D}/\partial t = -(2i\mu/\hbar)E(\mathcal{P} - \mathcal{P}^*) \tag{1.14}$$

Thus the assumption (1.7) has had the result of truncating an infinite hierarchy of moments, reducing it to the first two moments.

1.2 The Maxwell–Bloch equations

In the previous section, we have derived coupled equations that describe the interaction between light and *stable* atoms. However, atomic levels are not stable and have only a finite lifetime when they interact with an electromagnetic field, except for the ground state. How should we modify the theory developed in Section 1.1 to take this natural decay process into account? There are many ways to answer this question. However, all the theories that have been developed until now are unsatisfactory and avoid, one way or the other, the real difficulty. The problem is that there is no equation describing unstable atomic states in the fundamental way that the Schrödinger equation describes stable atomic states. However, we observe that practically all theories developed to handle this problem produce the same result in first approximation. We will therefore deal with this in a rather offhand manner. Two processes have to be added to equations (1.12)–(1.14) before we can use them to describe a laser and, more generally, cavity nonlinear optics.

1.2.1 Decay processes

The finite lifetime of excited atoms interacting with an electromagnetic field results in the decay of the atomic polarization $\mathcal{P}$ and of the population inversion $\mathcal{D}$. In the absence of atom–field interaction, $\mathcal{P}$ oscillates at the frequency ω_a and the population inversion $\mathcal{D}$ remains constant. A simple way to introduce phenomenologically the atomic decay process is to add to the equations describing the time evolution of $\mathcal{P}$ and $\mathcal{D}$ linear decay rates

$$\partial \mathcal{P}/\partial t = -i\omega_a \mathcal{P} - (i\mu/\hbar)E\mathcal{D} - \gamma_\perp \mathcal{P} \qquad (1.15)$$

$$\partial \mathcal{D}/\partial t = -(2i\mu/\hbar)E(\mathcal{P} - \mathcal{P}^*) - \gamma_\parallel \mathcal{D} \qquad (1.16)$$

The introduction of two different decay rates is justified by the fact that the macroscopic atomic polarization decays for yet another reason. The polarization of each atom is an oscillating function of the form $\mathcal{P}_k \sim \exp(i\omega_a t + i\varphi_k)$ in the absence of the atomic decay. The index k labels the atoms. If the phases are randomly distributed, the macroscopic polarization averages to zero. To assume that initially there is a nonvanishing macroscopic polarization requires a coherence in the form of phase relations among the atomic polarizations. During the evolution of the macroscopic polarization, these phase relations disappear as a result of the interaction of the excited atoms with their surroundings. This provides the second mechanism that causes the decay of the polarization.

1.2.2 Incoherent optical pumping

If atomic decay is unavoidable, incoherent excitation is unavoidable as well. Thermal radiation, for instance, excites atoms and molecules and produces the Planck distribution. Another example is the laser, a device in which there is **L**ight **A**mplification by **S**timulated **E**mission of **R**adiation. In order for the radiation to be amplified, it has to overcome the absorption processes that take place in the nonlinear medium. This can be achieved in many ways. The most popular is by pumping with an incoherent source the atoms into the upper state. We take the short route again and introduce, phenomenologically, the effect of pumping as a source of population inversion. This leads to the modification of the equation for $\mathcal{D}$ by the addition of a constant

$$\partial \mathcal{D}/\partial t = -(2i\mu/\hbar)E(\mathcal{P} - \mathcal{P}^*) - \gamma_\parallel(\mathcal{D} - \mathcal{D}_a) \qquad (1.17)$$

As a result, if we set $E = 0$, the material equations have solutions $(\mathcal{P}, \mathcal{D}) \rightarrow (0, \mathcal{D}_a)$ in the long time limit.

1.3 Complex amplitude equations

At this point, the atom–field system is described by the set of equations

$$c^2 \partial^2 E/\partial z^2 - \partial^2 E/\partial t^2 = (N\mu/\varepsilon_0) \partial^2(\mathcal{P} + \mathcal{P}^*)/\partial t^2 \qquad (1.18)$$

$$\partial \mathcal{P}/\partial t = -i\omega_a \mathcal{P} - (i\mu/\hbar)E\mathcal{D} - \gamma_\perp \mathcal{P} \qquad (1.19)$$

$$\partial \mathcal{D}/\partial t = -(2i\mu/\hbar)E(\mathcal{P} - \mathcal{P}^*) - \gamma_\parallel(\mathcal{D} - \mathcal{D}_a) \qquad (1.20)$$

Because we are interested in optical fields that oscillate in the visible (that is, with frequencies of the order of 10^{15} Hz), it is natural to separate in the electric field and the atomic polarization a fast oscillating term

$$E(z, t) = (1/2)\{E_0(z, t) \exp[i(k_c z - \omega_c t)]$$

$$+ E_0^*(z, t) \exp[-i(k_c z - \omega_c t)]\} \qquad (1.21)$$

$$\mathcal{P}(z, t) = (i/2)P_0(z, t) \exp[i(k_c z - \omega_c t)] \qquad (1.22)$$

The decomposition (1.21) and (1.22) is arbitrary as long as k_c and ω_c are not specified. The natural choice in optics is to select for k_c and ω_c those values assumed by the electric field in the absence of matter in the cavity. Thus $\omega_c^2 = (ck_c)^2 = (2\pi nc/L)^2$, $n = \pm 1, \pm 2, \ldots$ in a ring cavity of length L and $k_c^2 = (\pi n/L)^2$, $n = \pm 1, \pm 2, \ldots$ in a Fabry–Pérot cavity of length L. The time dependence of the complex amplitudes E_0 and P_0 is normally ruled by the atomic decay rates $\gamma_\perp$ and $\gamma_\parallel$ that are much smaller than the optical frequencies. Typically, $\gamma_\perp$ and $\gamma_\parallel$ are in the range 10^3 to 10^{11} Hz. Therefore the functions E_0, P_0, and $\mathcal{D}$ are slowly varying functions of time and space. The two material variables satisfy the equations

$$\partial P_0/\partial t = -i(\omega_a - \omega_c)P_0 - (\mu/\hbar)(E_0 + E_0^* e^{-2i\varphi})\mathcal{D} - \gamma_\perp P_0 \quad (1.23)$$

$$\partial \mathcal{D}/\partial t = (\mu/2\hbar)(P_0^* E_0 + P_0 E_0^* + E_0^* P_0^* e^{-2i\varphi}$$

$$+ E_0 P_0 e^{2i\varphi}) - \gamma_\parallel(\mathcal{D} - \mathcal{D}_a) \qquad (1.24)$$

where $\varphi \equiv k_c z - \omega_c t$. In both material equations, we find in the right-hand side resonant contributions, namely the autonomous terms and corrections oscillating at the frequency $2\omega_c$. Because we are dealing with optical frequencies, it is assumed that the correction terms do not contribute to the dynamical evolution of the system because they oscillate too fast and average to zero. Hence the Maxwell–Bloch equations become

$$c\,\partial E_0/\partial z + \partial E_0/\partial t = -(N\mu\omega_c/2\varepsilon_0)P_0$$

$$- (iN\mu/\varepsilon_0)\,\partial P_0/\partial t + (N\mu/2\varepsilon_0\omega_c)\,\partial^2 P_0/\partial t^2$$

$$- (i/2\omega_c)\,\partial^2 E_0/\partial t^2 + (ic^2/2\omega_c)\,\partial^2 E_0/\partial z^2 \quad (1.25)$$

$$\partial P_0/\partial t = -i(\omega_a - \omega_c)P_0 - (\mu/\hbar)E_0\mathcal{D} - \gamma_\perp P_0 \qquad (1.26)$$

$$\partial \mathcal{D}/\partial t = (\mu/2\hbar)(P_0^* E_0 + P_0 E_0^*) - \gamma_\parallel(\mathcal{D} - \mathcal{D}_a) \qquad (1.27)$$

The usual way to further simplify these equations is to introduce phenomenological linear cavity losses γ_c to the field equation, (1.25), and to express the fact that the envelopes E_0 and P_0 are slowly varying functions by imposing the inequalities

$$\omega_c|P_0| \gg |\partial P_0/\partial t|, \qquad \omega_c|E_0| \gg |\partial E_0/\partial t|, \qquad k_c|E_0| \gg |\partial E_0/\partial z| \quad (1.28)$$

This leads to the field equation

$$c\,\partial E_0/\partial z + \partial E_0/\partial t = -(N\mu\omega_c/2\varepsilon_0)P_0 - \gamma_c E_0 \tag{1.29}$$

In later chapters, we will use this "fast physics" approach. However, in this chapter we are going to justify field equation (1.29) in a more rigorous way based on an asymptotic analysis that includes the boundary conditions. The purpose of such an attitude is, first, to give a proper foundation to field equation (1.29) and, second, to determine the limits of its validity.

Considering again the set (1.25)–(1.27), we must add the boundary conditions for the field E_0. In most of this book, we consider ring lasers. Such lasers usually have an in-coupling mirror, an out-coupling mirror, and additional mirrors to close the optical path. The nonlinear medium is placed between the in-coupling and the out-coupling mirrors. We assume that the two coupling mirrors have the same reflectivity R and that the additional mirrors are perfect mirrors with $R = 1$. In such a configuration, the light follows a closed path in the cavity and therefore periodic boundaries are imposed: $E[0, t] = RE[\ell, t - (L - \ell)/c]$. This relation expresses the fact that the cavity has a length L, whereas the active medium fills only a length ℓ of the total cavity. Using the relation $ck_c = \omega_c$ leads to a similar relation for the complex, slowly varying field amplitude

$$E_0[0, t] = RE_0[\ell, t - (L - \ell)/c] \tag{1.30}$$

This boundary condition is not isochronous and therefore represents a barrier to the analysis of (1.25). Indeed, if we try to solve (1.25) by a Fourier series, no single Fourier component satisfies the boundary condition (1.30). This makes Fourier series useless for analyzing this problem. One way to get around this difficulty is to introduce the change of variables

$$z' = z,\, t' = t + (L - \ell)z/\ell c \tag{1.31}$$

We also introduce reduced dependent variables

$$E_0(z, t) = E(z', t')(\hbar/\mu)\sqrt{\gamma_\perp\gamma_\parallel}\exp[-z'\ln(R)/\ell] \tag{1.32}$$

$$P_0(z, t) = P(z', t')\mathcal{D}_a\sqrt{\gamma_\parallel/\gamma_\perp}\exp[-z'\ln(R)/\ell] \tag{1.33}$$

$$\mathcal{D}(z, t) = D(z', t')\mathcal{D}_a \tag{1.34}$$

and dimensionless space–time variables and parameters that will simplify the notation

$$\kappa = c|\ln(R)|/L\gamma_\perp, \qquad \alpha = \mathcal{D}_a\mu^2 N\omega_c/2\hbar\gamma_\perp c\varepsilon_0, \qquad A = \alpha\ell/|\ln(R)|$$

$$\tau = \gamma_\perp t', \quad v = c\ell/L\gamma_\perp, \quad \omega = \omega_c/\gamma_\perp, \quad \delta_{ac} = (\omega_a - \omega_c)/\gamma_\perp,$$

$$\gamma = \gamma_\parallel/\gamma_\perp \tag{1.35}$$

With this new notation, equations (1.25)–(1.27) become

$$\partial E/\partial \tau + v\, \partial E/\partial \zeta = -\kappa[E + AP + (2iA/\omega^2)\partial P/\partial \tau - (A/\omega^2)\partial^2 P/\partial \tau^2]$$

$$+ (iL\kappa^2/2\ell\omega)E + i[(L - 2\ell)/2\ell\omega]\partial^2 E/\partial \tau^2 + i[\kappa(L - \ell)/\ell\omega]\partial E/\partial \tau$$

$$+ i(Lv^2/2\ell\omega)\partial^2 E/\partial \zeta^2 + i[(L - \ell)v/\ell\omega]\partial E/\partial \tau\, \partial \zeta$$

$$+ i(\kappa Lv/\ell\omega)\partial E/\partial \zeta \tag{1.36}$$

$$\partial P/\partial \tau = -(1 + i\,\delta_{ac})P - ED \tag{1.37}$$

$$\partial D/\partial \tau = \gamma\left[1 - D + (1/2)(E^*P + EP^*)\exp\left[2\zeta|\ln(R)|/\ell\right]\right] \tag{1.38}$$

where $\zeta = z'$. Now the boundary condition is isochronous

$$E(0, \tau) = E(\ell, \tau)$$

At this point, we are in a good position to introduce two main approximations that lead to the usual semiclassical laser equations.

Uniform field limit. This limit, also known as the *mean field limit*, is based on the assumption that the medium is nearly optically transparent ($\alpha\ell \ll 1$) but the coupling mirrors are nearly perfect ($R \lesssim 1$) so that there is eventually a finite gain. These conditions are expressed in the form

$$-\ln(R) \to 0, \qquad \alpha\ell \to 0, \qquad A \equiv -\alpha\ell/\ln(R) = \mathcal{O}(1) \tag{1.39}$$

Slowly varying envelope approximation (SVEA). Lasers are characterized by the inequality $\omega \equiv \omega_c/\gamma_\perp \gg 1$ expressing the fact that the optical oscillation and the atomic relaxation take place on widely separated scales. Furthermore, the ratio ℓ/L is less or equal to unity. Two cases can be considered. Either ℓ/L is $\mathcal{O}(1)$ or it is a small parameter. In the latter case, we must specify how small is the ratio ℓ/L, in terms of the parameter $1/\omega$. Two examples of limits are

$$1/\omega = \varepsilon \ll 1, \qquad \ell/L = \mathcal{O}(1), \qquad \beta = L/\ell\omega = \mathcal{O}(\varepsilon) \tag{1.40}$$

$$1/\omega = \varepsilon \ll 1, \qquad \ell/L = \mathcal{O}(\varepsilon), \qquad \beta = L/\ell\omega = \mathcal{O}(1) \tag{1.41}$$

The natural limit for conventional lasers is (1.40). In the double limit (1.39) and (1.40), the Maxwell–Bloch equations become

$$\partial E/\partial \tau + v\, \partial E/\partial \zeta = -\kappa(E + AP) \tag{1.42}$$

$$\partial P/\partial \tau = -(1 + i\,\delta_{ac})P - ED \tag{1.43}$$

$$\partial D/\partial \tau = \gamma[1 - D + (1/2)(E^*P + EP^*)] \tag{1.44}$$

No restriction has been set on v since it can be absorbed in a rescaling of the space variable ζ. Note that owing to the rescaling of the variables and parameters, the pumping rate appears as the coefficient A in the field equation, whereas in the unscaled form of the equations the pumping rate $\mathcal{D}_a$ appears in the equation that governs the dynamics of the population inversion.

The other alternative represented by the limit (1.41) describes a laser with an extremely thin active medium in a very long cavity. Such a laser would have rather different properties from the usual laser, some of which having been investigated in Reference [5].

Equations (1.42)–(1.44) contain a bias. The choice of the decomposition (1.21) and (1.22) in terms of the propagating (or running) waves $\exp[i(k_c z - \omega_c t)]$ was motivated by the fact that they are eigenfunctions of the d'Alembert operator $c^2 \partial^2/\partial z^2 - \partial^2/\partial t^2$. However, the counterpropagating waves $\exp[i(k_c z + \omega_c t)]$ are also eigenfunctions of this operator and should have been included in the general decomposition (1.21)–(1.22). This leads to three types of solutions: running waves in any of the two directions and standing waves. The valid way to analyze the problem is to introduce a general decomposition of the form

$$
\begin{aligned}
E(z, t) = \{&E_1(z, t) \exp[i(k_c z - \omega_c t)] \\
&+ E_2(z, t) \exp[i(k_c z + \omega_c t)] + c.c.\}/2
\end{aligned}
\tag{1.45}
$$

$$
\begin{aligned}
\mathcal{P}(z, t) = i\{&P_1(z, t) \exp[i(k_c z - \omega_c t)] \\
&+ P_2(z, t) \exp[i(k_c z + \omega_c t)]\}/2
\end{aligned}
\tag{1.46}
$$

(where *c.c.* stands for *complex conjugate*) and to prove that the nonlinear equations for the ring cavity laser admit only unidirectional stable solutions. This problem has never been solved in its full extent. A partial answer is given in Chapter 9, where it is shown that on resonance (that is, for $\omega_c = \omega_a$), in steady state and for a ring cavity with identical losses in both directions, the standing wave is unstable and the two running waves are equally stable for space-independent field amplitudes. This means that either $E_1 = P_1 = 0$ or $E_2 = P_2 = 0$. Only the initial condition or fluctuations determine which of the two directions is taken by the field. This result has been extended to cover the detuned cavity in Reference [6], where a large bibliography can be found. This answer is only partial because the standing wave pattern might still be stabilized with a space–time-dependent amplitude. Another scenario is that an unstable time-dependent solution emerging from that solution becomes stable for larger values of the control parameter. Although this is not common, examples of this situation have been reported in nonlinear optics [7].

1.4 The single-mode laser

One of the most useful properties of a laser is that it can produce a quasi-monochromatic radiation field. That is, the laser field has a dispersion in frequency that can be made practically as small as one wishes. In that case, we can further reduce (1.42)–(1.44) by assuming that the electric field and the atomic polarization no longer depend on the space variable ξ

$$\begin{pmatrix} E(\zeta, \tau) \\ P(\zeta, \tau) \end{pmatrix} = \begin{pmatrix} E(\tau) \\ P(\tau) \end{pmatrix} \exp(-i\bar{\omega}\tau) \tag{1.47}$$

The choice of $\bar{\omega}$ is still arbitrary. We choose $\bar{\omega}$ to be the difference between the lasing frequency Ω (which is still unknown) and the cavity frequency ω_c. In this case, equations (1.42)–(1.44) reduce to one of the standard form of the single-mode laser equations

$$E' = -\kappa[(1 + i\Delta)E + AP] \tag{1.48}$$

$$P' = -(1 + i\delta)P - ED \tag{1.49}$$

$$D' = \gamma[1 - D + (1/2)(E^*P + EP^*)] \tag{1.50}$$

where the prime symbol ($'$) indicates a time derivative and the two detuning functions are defined as

$$\Delta = (\omega_c - \Omega)/\gamma_c, \qquad \delta = (\omega_a - \Omega)/\gamma_\perp, \qquad \gamma_c = \kappa\gamma_\perp \tag{1.51}$$

The detuning functions Δ and δ depend on the choice of the rotating reference frame in which the field E is described. Although this choice is to a large extent arbitrary, different physical phenomena are better described in different reference frames. The reader is referred to Reference [8] for a discussion of this point. One advantage of the choice (1.51) is that the electric field E has a steady state value since Ω is the lasing frequency. Indeed, equations (1.48)–(1.50) have steady state (or fixed points) solutions determined by $E' = P' = D' = 0$. The resulting algebraic equations have two solutions

$$E_s = 0, \qquad \Omega \text{ undetermined} \tag{1.52}$$

and

$$|E_s|^2 = A - 1 - \delta^2 \tag{1.53}$$

$$P_s = -E_s D_s/(1 + i\delta) \tag{1.54}$$

$$D_s = (1 + \delta^2)/(1 + \delta^2 + |E_s|^2) \tag{1.55}$$

$$\Delta + \delta = 0 \tag{1.56}$$

where the index s stands for steady solution. Equation (1.56) is an algebraic equation for the lasing frequency Ω. It is called the *dispersion relation,* and its solution can be written as

$$\begin{aligned} \Omega &= (\gamma_c\omega_a + \gamma_\perp\omega_c)/(\gamma_c + \gamma_\perp) \\ &= \omega_a + \gamma_\perp(\omega_c - \omega_a)/(\gamma_c + \gamma_\perp) \\ &= \omega_c + \gamma_c(\omega_a - \omega_c)/(\gamma_c + \gamma_\perp) \end{aligned} \tag{1.57}$$

Thus the lasing frequency always lies between the atomic and the cavity frequencies.

Using Equation (1.56), which defines the lasing frequency, we further simplify the single-mode laser equations

$$E' = -\kappa[(1 + i\,\Delta)E + AP] \tag{1.58}$$

$$P' = -(1 - i\,\Delta)P - ED \tag{1.59}$$

$$D' = \gamma[1 - D + (1/2)(E^*P + EP^*)] \tag{1.60}$$

where $\Delta = (\omega_c - \omega_a)/(\gamma_c + \gamma_\perp)$ is a well-defined parameter independent of the pump parameter A.

1.5 Linear stability analysis

The nonlinearity of the laser equations (1.58)–(1.60) induces two simultaneous steady solutions, the trivial solution (1.52) and the nontrivial solution (1.53)–(1.56). Each of these solutions has its basin of attraction, that is, a set of initial conditions from which the system evolves toward the steady state. The determination of the basin of attraction is, in most cases, impossible to carry out analytically. Even numerically, it is a major endeavor that has rarely been conducted successfully. However, a more modest question can be asked, namely: What is the stability of a steady state solution when the system is infinitesimally perturbed? This leads to a *linear* stability analysis. To see how it proceeds, we assume that the system is initially in the trivial state

$$E_s = P_s = 0, \qquad D_s = 1 \tag{1.61}$$

To test the stability of that solution against infinitesimal perturbations amounts to constructing solutions of laser equations (1.58)–(1.60) of the form

$$E = E_s + \varepsilon e + \mathcal{O}(\varepsilon^2), \qquad P = P_s + \varepsilon p + \mathcal{O}(\varepsilon^2),$$
$$D = D_s + \varepsilon d + \mathcal{O}(\varepsilon^2) \tag{1.62}$$

where ε is a small parameter and e, p, and d are $\mathcal{O}(1)$ functions. These expressions are inserted into the laser equations. The right-hand side of these equations contains terms proportional to ε and ε^2. Because we neglected the $\mathcal{O}(\varepsilon^2)$ terms in (1.62), by consistency we also neglect them in the evolution equations. As a result, the laser equations become linear homogeneous equations for the deviations, with constant coefficient. A fundamental theorem in the theory of ordinary differential equations guarantees (under rather general conditions that are always true here) that the solution is a sum of exponentials of the form $\exp(\lambda_k t)$. There are as many roots λ_k as there are differential equations. The beautiful property is that the λ_k are the roots of a polynomial of degree N, the characteristic equation, where N is the number of differential equations. In the case of the trivial solution (1.61), this procedure is easy to carry through. The linearized equations for the deviations are

$$e' = -\kappa[(1 + i\,\Delta)e + Ap] \tag{1.63}$$

$$p' = -(1 - i\,\Delta)p - e \tag{1.64}$$

$$d' = -\gamma d \tag{1.65}$$

The last equation tells us that $d \to 0$ as $t \to \infty$. The first two equations have solutions of the form

$$\mathbf{u} \equiv \begin{pmatrix} e \\ p \end{pmatrix} = \begin{pmatrix} e_1 \\ p_1 \end{pmatrix} \exp(\lambda_1 t) + \begin{pmatrix} e_2 \\ p_2 \end{pmatrix} \exp(\lambda_2 t) \tag{1.66}$$

where λ_1 and λ_2 are the two roots of the quadratic equation

$$\lambda^2 + \lambda[\kappa(1 + i\,\Delta) + 1 - i\,\Delta] + \kappa(1 + \Delta^2 - A) = 0 \tag{1.67}$$

The two roots are complex. Their sum has negative real part $-\kappa - 1$. Because their product is $\kappa(1 + \Delta^2 - A)$, both roots have a negative real part if $A < 1 + \Delta^2$. In that case, $\mathbf{u} \to \mathbf{0}$ as $t \to \infty$. This result means that an infinitesimal perturbation of the solution (1.61) decreases exponentially in time and the solution (1.61) is called asymptotically stable for $A < 1 + \Delta^2$. Conversely, if $A > 1 + \Delta^2$, one root has a positive real part and the solution (1.66) diverges in time. In that case, the solution (1.61) is unstable against infinitesimal perturbations.

The same analysis can be repeated for the nontrivial solution

$$|E_s|^2 = A - 1 - \Delta^2 \tag{1.68}$$

$$P_s = -E_s D_s / (1 - i\,\Delta) \tag{1.69}$$

$$D_s = (1 + \Delta^2)/(1 + \Delta^2 + |E_s|^2) \tag{1.70}$$

Using expansion (1.62) with this steady state, there is no more decoupling of the population inversion from the other two variables because E_s and P_s do not vanish. Hence the problem becomes five-dimensional and the characteristic equation for λ is the quintic

$$P(5, \lambda) = \lambda \sum_{n=1}^{4} a_n \lambda^n = 0 \tag{1.71}$$

$$a_4 = 1$$

$$a_3 = \gamma + 2(\kappa + 1)$$

$$a_2 = 2\gamma(1 + \kappa) + (1 + \kappa)^2 + (A - 1 - \Delta^2)\gamma + (1 - \kappa)^2 \Delta^2$$

$$a_1 = \gamma[(1 + \kappa)^2 + (1 + 3\kappa)(A - 1 - \Delta^2) + (1 - \kappa)^2 \Delta^2]$$

$$a_0 = 2\gamma\kappa(1 + \kappa)(A - 1 - \Delta^2)$$

A comprehensive discussion of this polynomial equation would lead us away from our main objective, which is to prepare the reader for the analyses of the later chapters. References [9] and [10] give a review of the properties of this polynomial and of the laser above the instability threshold of the steady lasing mode. However, a number of simple results can be easily inferred from (1.71): (1) Among the five roots of $P(5, \lambda) = 0$, one root is always zero. This corresponds to the fact that E, and P being complex, only their phase difference is a variable with dynamical relevance for the evolution of the amplitude variables. If we add to each phase an arbitrary function of time (corresponding to a time-dependent rotation frequency of the reference frame), the dynamics of the amplitudes will not be changed. This invariance finds its expression in the vanishing root. (2) The product of the remaining four roots equals a_0. A change of stability is expected when the sign of a_0 changes; that is, if $A = A_c = 1 + \Delta^2$. No surprise: This is precisely the value of the control parameter A where the trivial solution (1.61) lost its stability. To obtain more information on the change of stability that occurs at $A = A_c$, we make a local analysis and introduce a vicinity of the critical point $A = A_c + \eta$ with $0 < \eta \ll 1$. It follows from (1.71) that three roots are $\mathcal{O}(1)$, and one root is proportional to the small parameter η and is given by

$$\lambda = -\frac{2\eta\kappa(\kappa + 1)}{(\kappa + 1)^2 + (\kappa - 1)^2 \Delta^2} + \mathcal{O}(\eta^2) < 0 \qquad (1.72)$$

Hence the nontrivial solution is stable if the remaining three roots have negative real parts, which can be shown without difficulty. The restriction $\eta > 0$ stems from the condition $|E_s|^2 > 0$. (3) At a finite distance of A_c, a second threshold occurs if $\kappa > 1 + \gamma$ [see (9.29) for an explicit expression of the pump parameter A and the intensity at the second threshold]. Beyond that second threshold, the behavior of the lasing intensity can be either periodic or chaotic, depending on the detuning. For small detuning (typically $\Delta^2 < 3$), the lasing intensity is chaotic, whereas for large detuning ($\Delta^2 > 3$) it is periodic.

This last section has given the elements of the laser stability theory that are necessary for the understanding of the next chapters. However, laser stability is an important topic and is studied more deeply in Chapter 9.

References

[1]   H. Haken, Laser Theory in *Encyclopedia of Physics*. L. Genzel, ed. (Springer, Heidelberg, 1970).

[2]   M. Sargent III, M. O. Scully, and W. Lamb, Jr., *Laser Physics* (Addison-Wesley, Reading, 1974).

[3]   H. Haken, *Light,* Vol. 1 (North-Holland, Amsterdam, 1981).
[4]   H. Haken, *Light,* Vol. 2 (North-Holland, Amsterdam, 1981).
[5]   R.-D. Li and P. Mandel, *Opt. Commun.* **75** (1990) 72.
[6]   H. Zeghlache, P. Mandel, N. B. Abraham, L. M. Hoffer, G. L. Lippi, and
        T. Mello, *Phys. Rev. A* **35** (1988) 470.
[7]   See Figure 1c in X. G. Wu and P. Mandel, *J. Opt. Soc. Am B* **3** (1986) 724
        and Figure 9.6 in Reference [9].
[8]   E. Roldán, G. J. de Valcárcel, R. Vilaseca, and P. Mandel, *Phys. Rev. A* **48**
        (1993) 591; V. Y. Toronov and V. L. Derbov, *Phys. Rev. A* **49** (1994) 1392.
[9]   N. B. Abraham, P. Mandel, and L. M. Narducci, Dynamical instabilities and
        pulsations in lasers in *Progress in Optics,* Vol. 25, pp 1–190, E. Wolf, ed.
        (North-Holland, Amsterdam, 1988).
[10]   A. A. Bakasov and N. B. Abraham, *Phys. Rev. A* **48** (1993) 1633.

2

Parameter swept across a steady bifurcation I

2.1 Introduction

At the end of the previous chapter, we have seen that a characteristic of non-linear systems is the presence of critical points, that is, values of the control parameter at which two solutions coincide. Note that at a critical point, more than two solutions can coexist and that the coexisting solutions need not be stationary: They can have any time dependence. In the next four chapters, though, we concentrate on the steady critical points at which two steady solutions coincide. If the solutions exist on both sides of the critical point, we call it a *bifurcation point*. Another type of critical point that will also draw much of our attention later is the limit point. However, the solutions exist on only one side of the critical point.[1]

The main feature of critical points is that their presence is always signaled by the vanishing of the real part of at least one characteristic root in the stability analysis. For steady critical points, it is a real root that vanishes. In some cases, however, more than one root will vanish at criticality. Then one deals with degenerate critical points that may have richer properties. Physically, the absolute value of the real part of a characteristic root λ is a relaxation rate if $\text{Re}(\lambda) < 0$ and a divergence rate if $\text{Re}(\lambda) > 0$. Let λ_c be the root that vanishes at the critical point. The vanishing of a relaxation rate at criticality means the divergence of a relaxation time. Thus, an unconventional dynamics occurs in the domain surrounding the critical point that is characterized by *critical slowing down*. Its main property is that the longest relaxation time (that is expected to control the dynamics) is none of the physical relaxation times, such as κ or γ in the laser problem, but rather the inverse of the characteristic root that vanishes at criticality. Thus, no matter how large the physical characteristic times are, there

---

[1] In general, the real solutions that exist on one side of the limit point become complex on the other side of the limit point. These complex solutions are usually not physically admissible, and therefore they are not even mentioned.

always exists a vicinity of the critical point where $1/\left|\text{Re}(\lambda_c)\right|$ is much larger than any of them. In that domain, the dynamics is universal in the sense that it is essentially independent of the details of the physical problem at hand. The dominant effects come from such considerations as the local geometry of the steady solutions that meet at the critical point.

One of the activities in theoretical physics is to set up models that represent experimental situations. Good modeling presupposes that the essential features of the system have been identified. However, the real test of a model is its compatibility with experimental results. This is by no means a simple task because quite often many additional simplifying assumptions have been introduced between the model formulation and the final result that is presented as a test of the model. Another difficulty is directly related to the occurrence of critical points. The vicinity of a critical point may be very difficult to study experimentally or numerically because of a bad case of critical slowing down. For instance, a relaxation time of two weeks has been reported in a study of the Bénard instability. The natural way to bypass this difficulty seems to sweep the control parameter across the critical point. If the sweep rate is small enough, intuition suggests that there should hardly be a difference between the static and the dynamic (or swept system) properties. In fact, in the limit of infinitely slow sweep, the system should follow adiabatically the steady state. The central theme of this book is that this view is wrong because it neglects critical slowing down.

2.2 Sweeping across the laser first threshold

2.2.1 General formulation

As a clear example of how wrong our intuition may be, let us consider again the laser equations and suppose that we want to study the vicinity of the bifurcation point $A = A_c = 1 + \Delta^2$ by sweeping the control parameter A. For the sake of argument, we simplify the model by assuming that the laser is on resonance, $\omega_a = \omega_c$, so that $\Delta = 0$. In that case, the laser equations take the form

$$dE/dt = -\kappa(E + \mathcal{A}P) \qquad (2.1)$$

$$dP/dt = -P - ED \qquad (2.2)$$

$$dD/dt = \gamma(1 - D + EP) \qquad (2.3)$$

where the variables E and P are chosen to be real. To simplify the notation, we use t for the reduced time that was notated as τ in Chapter 1. The difference with the problem discussed in Chapter 1 is that the optical pump parameter $\mathcal{A}$ depends on time. This transforms the laser equations, (2.1)–(2.3), into a system of nonautonomous differential equations. Contrary to the autonomous equations associated with a constant A, the nonautonomous equations usually do not have a steady state and at any time the system keeps the memory of its initial condition and all its previous evolution.

As a first approach to this problem, we assume that A depends on a slow time εt with $0 < \varepsilon \ll 1$ such that the sweep rate ε is much smaller than the physical decay rates κ and γ. Because we wish to sweep the pump parameter across the critical point $A_c = 1$, we also impose $A(0) < 1$. The first result that is easy to verify is that even if A is a time-dependent function, the trivial solution $(E, P, D) = (0, 0, 1)$ remains an exact solution of the equations (2.1)–(2.3). This fortunate situation suggests a linear stability analysis similar to that of Section 1.5. Assuming solutions of the form

$$E = \eta e + \mathcal{O}(\eta^2), \quad P = \eta p + \mathcal{O}(\eta^2), \quad D = 1 + \eta d + \mathcal{O}(\eta^2)$$
$$0 < \eta \ll 1 \tag{2.4}$$

we obtain the linearized evolution equations

$$de/dt \equiv e' = -\kappa(e + Ap) \tag{2.5}$$
$$p' = -p - e \tag{2.6}$$

For consistency, we impose that the expansion (2.4) also holds for the initial condition. In this way, we describe the evolution of the system as long as it remains in the vicinity of the trivial solution. From (2.5)–(2.6) we obtain a closed equation for e

$$e'' + (\kappa + 1)e' + \kappa(1 - A)e = A'(e' + \kappa e)/A \tag{2.7}$$

We seek solutions of (2.7) that are similar to the solutions (1.66) of the corresponding autonomous system

$$e(\varepsilon, \varepsilon t) = c_1 \exp\left[\frac{1}{\varepsilon}\int_0^{\varepsilon t} g_1(\varepsilon, \tau)\, d\tau\right] + c_2 \exp\left[\frac{1}{\varepsilon}\int_0^{\varepsilon t} g_2(\varepsilon, \tau)\, d\tau\right] \tag{2.8}$$

The functions g_j satisfy the differential equation

$$g^2 + (\kappa + 1)g + \kappa(1 - A)g = \varepsilon\left[\frac{1 + g}{A}\frac{dA}{d\tau} - \frac{dg}{d\tau}\right], \qquad \tau = \varepsilon t \tag{2.9}$$

Expanding the functions g in power series of ε, $g(\varepsilon, \tau) = h(\tau) + \mathcal{O}(\varepsilon)$, we obtain for the dominant order contribution

$$h^2 + (\kappa + 1)h + \kappa(1 - A) = 0 \tag{2.10}$$

This is precisely the result (1.67) obtained for the autonomous problem, with two restrictions, however: First, in this case, the solution h depends on the slow time τ via $\mathcal{A}$. Second, because we are dealing with a nonautonomous problem, the function h that resembles so closely its autonomous counterpart (1.67) with $\Delta = 0$ is only an approximation since we have neglected the right-hand side of (2.9). From the analogy between the two equations, we conclude without further analysis that one root of (2.10), say h_1, is always negative and the other root, h_2, is negative for $\mathcal{A} < 1$ and positive for $\mathcal{A} > 1$. However, the positivity of h is not sufficient to destabilize the linearized solution (2.8). In analogy with the static case where the threshold for instability is defined by the vanishing of the argument in the exponential, we define a condition of instability in the nonautonomous case by the same criterion, that is

$$\int_0^{t^*} h_2(s)\, ds = 0 \qquad (2.11)$$

This is an implicit equation for the critical time t^* at which the solution begins to diverge. Although this definition is convenient from a theoretical viewpoint, the experimental determination of t^* may be a challenge. Indeed, in the domain that follows $t = t^*$, the increase of the field amplitude or the field intensity $|E|^2$ is exponential because usually there will be a rather steep jump toward another attractor. The steepness of this jump, which increases as ε decreases, may be a source of trouble for a clear-cut measurement of t^*. Nevertheless, the critical time t^* as defined by (2.11) is a useful concept and can be thought of as a lower bound to the measurable critical time. Another time that may be defined is the intermediate time $\bar{t}$ at which the static bifurcation is reached: $\mathcal{A}(\bar{t}) \equiv 1$. In terms of this time, the instability condition becomes

$$\int_0^{\bar{t}} h_2(s)\, ds = \int_{\bar{t}}^{t^*} \left[-h_2(s) \right]\, ds \qquad (2.12)$$

This relation expresses a balance between the stability accumulated in the first part of the sweep, where $0 \le t \le \bar{t}$ and $h_2 > 0$, and the instability accumulated in the second part of the sweep, where $\bar{t} \le t \le t^*$ and $h_2 < 0$. It is already quite obvious at this stage of the discussion that $t^* \ge \bar{t}$: The dynamical bifurcation is shifted with respect to the static bifurcation. Equivalently, it can be said that the sweep has induced a dynamical stabilization of the nonlasing solution.

2.2.2 Linear sweep

To get more insight into the amplitude of the delay suffered by the bifurcation because of the sweep, let us specify the nature of the sweep. The simplest choice is a linear sweep $\mathcal{A}(\varepsilon t) = a + \varepsilon t$. Solving (2.11) for t^* and using that

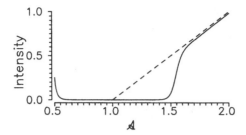

Figure 2.1 Equations (2.1)–(2.3) are integrated to obtain the intensity E^2 versus the gain parameter $\mathcal{A} = a + \varepsilon t$ (solid line) compared with the function $\mathcal{A} - 1$ (dotted line). Parameters are $\varepsilon = 0.001$, $\gamma = 1$, and $\kappa = 0.1$. Initial condition: $E(0) = 0.5$, $P(0) = 0$, $D(0) = 1$, and $a = 0.5$.

result to define the relative delay $\mathcal{D} = [\mathcal{A}(t^*) - \mathcal{A}(\bar{t})]/[\mathcal{A}(\bar{t}) - \mathcal{A}(0)] \equiv [\mathcal{A}^* - 1]/[1 - a]$ leads to

$$\mathcal{D} = \frac{1}{8\alpha}\left[4\alpha - 3 + 4\sqrt{-3\alpha^2 + \frac{39}{2}\alpha - \frac{183}{16} + 12(1-\alpha)^{3/2}}\right] \quad (2.13)$$

where $\alpha = 4\kappa(1 - a)/(1 + \kappa^2)$. This result is rather surprising and needs some comments.

1. $\mathcal{D}$ depends on the single variable α.

2. $\mathcal{D}$ varies from 1 to 5/4 as α varies from 0 to 1. This implies that the difference $\mathcal{A}(t^*) - \mathcal{A}(\bar{t})$ is greater than, or at least equal to, the difference $\mathcal{A}(\bar{t}) - \mathcal{A}(0)$. This property is clearly displayed in Figure 2.1.

3. The variable α is invariant for a change of κ into $1/\kappa$: The delay is not a good or a bad cavity property.

4. The most surprising property of the relative delay is its independence of the sweep rate ε.

This latter property is key to understanding the delay phenomenon. Indeed, it tells us that the limit $\varepsilon \to 0$ is singular since $\mathcal{D}$ is undefined if $\varepsilon = 0$ and $\mathcal{D} = \mathcal{O}(1)$ for arbitrarily small sweep rates. This singularity was already apparent in the solution (2.8). It is also displayed in Figure 2.2 where it is clear that the limit $\varepsilon \to 0$ does not coincide with the function $\mathcal{A} - 1$: The delay remains, as determined by the linearized result (2.13), but the jump transition to the upper branch becomes steeper.

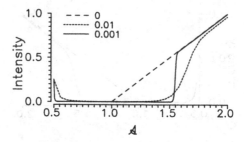

Figure 2.2 Influence of the sweep rate on the time evolution of the intensity obtained by solving equations (2.1)–(2.3) with $\mathcal{A} = a + \varepsilon t$. Parameters are $\gamma = 1$ and $\kappa = 0.5$. Initial condition: $E(0) = 0.5$, $P(0) = 0$, $D(0) = 1$, and $a = 0.5$. The sweep rates are 0.01 and 0.001. The function $\mathcal{A} - 1$ is plotted for reference.

In fact, physically, it is easy to understand (with hindsight!) that there is a qualitative barrier between the static situation and the dynamic situation in which the control parameter varies in time. No matter how slowly the control parameter is swept, if the sweep rate is not zero there is no limit in which the small sweep rate can reproduce the properties of the static case. However, we will see later in this book that under some conditions, the slow sweep can mimic some aspects of the static problem.

2.2.3 Exact solution of the linearized problem

Most results of the previous two sections were obtained by solving asymptotically the linearized equations (2.5) and (2.6). In particular, the instability condition (2.11) results from an expansion in powers of ε of the general solution g that satisfies (2.9). To make sure that none of the above results depends critically on these approximations, we briefly discuss the exact solution of (2.5) and (2.6). The simplest way to obtain the exact solution is by eliminating e from the linearized equations to obtain a closed equation for p. This leads to $p'' + (\kappa + 1)p' + \kappa(1 - \mathcal{A})p = 0$. Introducing $p = q \exp[-(\kappa + 1)t/2]$ yields $q'' + [-(\kappa + 1)^2/4 + \kappa(1 - \mathcal{A})]q = 0$. For $\mathcal{A} = a + vt$, this equation can be reduced to the Airy equation $d^2q/d\xi^2 - \xi q = 0$ where $\xi = (\kappa v)^{-2/3}[(\kappa + 1)^2/4 - \kappa(1 - \mathcal{A})]$. Thus the general solution of the linearized equations is of the form

$$p(t) = [aAi(\xi) + bBi(\xi)]\exp[-(\kappa + 1)t/2] \qquad (2.14)$$

where a and b are constants determined by the initial conditions and $Ai'(\xi) \equiv dAi(\xi)/d\xi$. For the initial condition, we take $e(0) = e_0$ and $p(0) = 0$ to simplify the results. Using the relation $Ai(\xi)Bi'(\xi) - Ai'(\xi)Bi(\xi) = \pi$ and $e(0) = -p(0) - p'(0)$, it follows that $a = -\pi e_0(\kappa v)^{-1/3} Bi(\xi_0)$ and $b = \pi e_0(\kappa v)^{-1/3} Ai(\xi_0)$ with $\xi_0 \equiv \xi(t = 0)$. This determines completely the atomic polarization

$$p(t) = \pi e_0(\kappa v)^{-1/3}[Bi(\xi)Ai(\xi_0) - Ai(\xi)Bi(\xi_0)] \exp[-(\kappa + 1)t/2] \quad (2.15)$$

From this expression, it is a simple matter to derive the solution for the field amplitude $e(t) = -p(t) - p'(t)$.

With the exact solution, we can define the critical time $\xi^* \equiv \xi(t^*)$ by the condition

$$\pi e_0(\kappa v)^{-1/3}[Bi(\xi^*)Ai(\xi_0) - Ai(\xi^*)Bi(\xi_0)] \exp[-(\kappa + 1)t^*/2] = 1 \quad (2.16)$$

This result has been derived for an arbitrary sweep rate v. To compare (2.16) with (2.13), we have to take the small sweep rate limit $v \to 0$. In that limit, $\xi \to +\infty$ and therefore we can use the well-known asymptotic expansions

$$Ai(\xi) \sim \frac{\exp(-\varphi)}{2\sqrt{\pi}\xi^{1/4}}[1 + \mathcal{O}(1/\varphi)], \quad Ai'(\xi) \sim -\frac{\xi^{1/4}\exp(-\varphi)}{2\sqrt{\pi}}[1 + \mathcal{O}(1/\varphi)]$$

$$Bi(\xi) \sim \frac{\exp(\varphi)}{\sqrt{\pi}\xi^{1/4}}[1 + \mathcal{O}(1/\varphi)], \quad Bi'(\xi) \sim -\frac{\xi^{1/4}\exp(\varphi)}{\sqrt{\pi}}[1 + \mathcal{O}(1/\varphi)]$$

$$(2.17)$$

where $\varphi = (2/3)\xi^{3/2}$. Inserting these expansions into (2.16) yields

$$\frac{2}{3}(\xi^*)^{2/3} - \frac{1}{4}\ln\xi^* - \frac{\kappa + 1}{2}t^* = \ln\frac{(\kappa v)^{1/3}}{\pi^{1/2}e_0 Ai(\xi_0)} \quad (2.18)$$

which, to leading order in the small parameter v, becomes

$$\frac{2}{3}(\xi^*)^{2/3} - \frac{\kappa + 1}{2}t^* \cong \frac{2}{3}(\xi_0)^{2/3} \quad (2.19)$$

With the definition of ξ_0 and of ξ^*, it is a matter of simple algebra to show that (2.19) leads to (2.13).

One result of the linearized theory is that it leaves no room for the parameter γ to play a role. A numerical integration of the nonlinear equations (2.1)–(2.3) shows that indeed γ starts to influence the dynamics of the transition *after* the intensity has left the vicinity of the trivial solution, that is, in the domain where the linearized approximation loses its validity. This is exemplified in Figure 2.3. The oscillations appearing in that figure as $\gamma \to 0$ are a signature of relaxation oscillations that occur in the bad cavity limit. Stated differently, a linear stability of the upper branch of solutions with constant pump parameter and in that domain of parameters produces a pair of complex roots, the imaginary part of

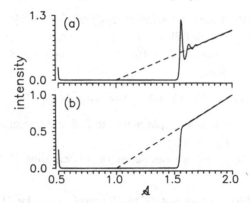

Figure 2.3 Influence of γ on the jump transition. The solid lines are obtained by integration of equations (2.1)–(2.3) with $\mathcal{A} = a + \varepsilon t$, whereas the dashed lines are the function $\mathcal{A} - 1$. Parameters are $\varepsilon = 0.001$ and $\kappa = 0.5$. Initial condition: $E(0) = 0.5$, $P(0) = 0$, $D(0) = 1$, and $a = 0.5$. (a): $\gamma = 0.1$; (b): $\gamma = 1$.

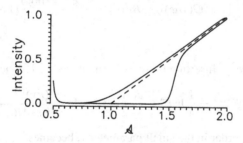

Figure 2.4 Forward and backward sweep. Same equations and parameters as in Figure 2.1. The sign of the sweep rate is changed at $\mathcal{A} = 2$. After that, $\mathcal{A}$ decreases to 0.6.

which is the relaxation oscillation frequency. These oscillations are probed as the gain parameter is swept across the dynamical bifurcation point at $\mathcal{A}(t^*)$. That is tantamount to subjecting the finite intensity solution to a strong perturbation.

Up to now, we have considered only forward sweeps. Let us change the sign of the sweep rate after the solution has moved to the vicinity of $\mathcal{A} - 1$. Figure 2.4 illustrates this situation. In the backward sweep there is no jump transition since the only domain in which there is critical slowing down is the vicinity of $\mathcal{A} = 1$ where, indeed, the dynamical solution departs somewhat from the function $\mathcal{A} - 1$.

2.2.4 Good cavity limit

Now that we have established the existence of the delay in one specific case, we wish to formulate the problem in a more generic way, one that does not rely any more on laser physics. In this way, we will be able to formulate general properties of the delayed bifurcation and also to extend our analysis. One way to achieve this goal is to start again from the Maxwell–Bloch equations (2.1)–(2.3) and to introduce the good cavity limit defined by the inequalities $\kappa \ll 1$ and $\gamma = \mathcal{O}(1)$. Thus it is assumed that the two atomic variables have characteristic decay rates much larger than the field decay rate. In that limit, we seek solutions of equations (2.1)–(2.3) in the form $Z(\kappa,t) = Z(\kappa,\tau) = Z_0(\tau) + \kappa Z_1(\tau) + \cdots$ where Z stands for any of the three variables E, P, and D and $\tau = \kappa t$ is a slow time. This leads, for the material variables, to the algebraic relations

$$D_0(\tau) = 1/[1 + E_0^2(\tau)], \qquad P_0(\tau) = -E_0(\tau)/[1 + E_0^2(\tau)] \qquad (2.20)$$

This procedure is also known as the *adiabatic elimination of the atomic variables*. It implies that at any given time, the atomic variables depend only on the instantaneous value of the electric field and not on its previous history. It is a kind of local stationarity assumption in the time domain, somewhat similar in its philosophy to the local equilibrium assumption in statistical mechanics.[2] Using the result (2.20), the field equation becomes

$$dE_0/dt = E_0[-\kappa + \kappa A/(1 + E_0^2)] \qquad (2.21)$$

The reason we have written this equation in terms of the unscaled time t is that κ is proportional to the cavity losses whereas the product κA is the linear gain of the field and does not depend on the cavity losses ($A \sim 1/\kappa$). Experimentally, both parameters can be controlled separately and can be swept in time independently. This represents two different classes of problems to analyze.

A further simplification of (2.21) is provided by the fact that we are mostly interested in the behavior of the system near its trivial solution $E_0 = 0$. Thus we expand (2.21) around that trivial state and obtain

$$dE_0/dt = E_0(-\kappa + \kappa A - \kappa A E_0^2) \qquad (2.22)$$

where we have retained the first nonlinear contribution.

To justify the consideration of such simple equations, we analyze them in the case of a gain sweep with A replaced by $\mathcal{A} = a + vt, v > 0$ and the initial condition $0 < E_0(0) \ll 1, a < 1$. The linearized equation is $dE_0/dt = \kappa E_0(-1 + a + vt)$, and its solution is $E_0(t) = E_0(0) \exp[(a - 1 + vt/2)\kappa t]$.

---

[2] In Chapter 12, we come back to this adiabatic procedure and we review it more critically.

Thus the instability condition is

$$t^* = 2(1 - a)/v \qquad (2.23)$$

which is equivalent to

$$\mathcal{A}^* - 1 = 1 - a \qquad \text{or} \qquad t^* = 2\bar{t} \qquad (2.24)$$

Hence the delay persists and has reached its lower bound. This indicates that by reducing the 3-D laser equations to a 1-D equation for the field, we have not lost any essential aspect of the problem. However, we have gained something because with the simpler equations, either (2.21) or (2.22), we are able to treat far more elaborate situations than would have been possible with the 3-D equations. The remainder of this book is devoted to analyzing approximations of the Maxwell–Bloch equations (and their generalization to cover optical bistability) in terms of 1-D equations. The exceptions will be Chapters 7 and 8, which deal with multimode lasers, and Chapter 9, whose purpose is precisely to show that retaining the full 5-D Maxwell–Bloch equations for a single-mode ring laser and the 9-D Maxwell–Bloch equations for counterpropagating modes in a ring laser does not prevent an analytic study of the laser stability.

What has become obvious by now is that the delay is independent of the direction of the bifurcating solution. That is, the delay is the same whether the steady state solution with finite amplitude bifurcates supercritically as in equation (2.22) or subcritically as in the equation

$$dE_0/dt = E_0(-\kappa + \kappa A + \kappa A E_0^2) \qquad (2.25)$$

This equation occurs, for instance, in the theory of a laser with saturable absorber close to the lasing threshold and the negative slope of E_0^2 versus A indicates a static hysteresis or bistability as shown in Figures 11.2 and 11.3. Hence a delay and a jump transition when the control parameter is swept in time can signal either a steady bifurcation or a static hysteresis. We demonstrate in Section 5.1 that the difference between these two situations is the dependence of the delay $\mathcal{D}$ on the sweep rate.

3

Parameter swept across a steady bifurcation II

In the previous chapter, we have justified the use of a 1-D approximate equation to study the delayed bifurcation if the sweep begins near the trivial steady state. However, this study was limited in a number of ways, mathematical and physical. In this chapter, we take the 1-D description for granted and extend the analysis. Since our approach to solve the delay problem is to use asymptotic methods, we have to be careful in specifying the order of magnitude of *all* parameters with respect to a given small parameter. The problems treated in this chapter involve at least two small parameters, one of which being always the sweep rate. We analyze how the asymptotics – and therefore the physical description of the system – is modified by an additional small parameter.

3.1 Influence of the initial condition

In this section, we use the 1-D formulation of the laser equations (2.21) that retains the full saturation

$$dE/dt = E\left[-\kappa + \kappa A/(1 + E^2)\right] \tag{3.1}$$

In the static case, this equation describes correctly the two steady states of the laser equations, namely the trivial state (laser off) and the nontrivial state (laser on). Thus by analyzing this equation with a time-dependent parameter, we expect to describe not only the delayed bifurcation but also the jump transition and the state of the laser after the jump. The experiments that were performed to test this theory used an Ar^+ laser that is well described by a 1-D equation such as (3.1). The experimental procedure was to vary in time the cavity losses [1].

Thus we analyze (3.1) with $-\kappa + \kappa A$ replaced by $-\kappa + \kappa_0 A$, where $\kappa = \kappa_0(1 - \bar{v}t)$, $0 < \bar{v} \ll 1$. We can absorb the initial loss κ_0 in a rescaling of time so that the equation to analyze becomes

$$dI/d\tau = 2I[-1 + v\tau + A/(1 + I)], \qquad 0 < v \ll 1 \tag{3.2}$$

with $\tau = \kappa_0 t$, $v = \bar{v}/\kappa_0$, and $I = |E|^2$ is interpreted as the field intensity. The linearization of this equation around $I = 0$ yields $dI/d\tau = 2I(A - 1 + v\tau)$, indicating that the two relevant parameters will be the sweep rate v and the deviation from the steady bifurcation $A - 1$. The initial condition that is referred to in the title of this section is the value of A. It is indeed an initial condition since it depends on the initial loss: $A \sim 1/\kappa_0$. The fact that it is the deviation from the static threshold $A - 1$ that matters may appear as a surprise because for the nonautonomous equation (3.2), $A = 1$ is not a critical point. This is precisely where the concept of initial condition becomes meaningful. Indeed, if $A > 1$, there is no delay; whereas if $A < 1$, we have seen that there is a delay. In the limiting case $A = 1$, the system is in a situation that separates the domain with no delay from the domain with an $\mathcal{O}(1)$ delay. Clearly, the vicinity of $A = 1$ must be some kind of boundary layer that bridges the two domains and in which different behavior must be expected. Therefore we study the two domains separately.

3.1.1 The regular problem

In this section we analyze (3.2) in the regular case $A - 1 = \mathcal{O}(1)$ with respect to the small parameter v and $I(0) = I_0$ with $0 < I_0 \ll 1$. We construct the solution of (3.2) in the three domains, $\tau < \tau^*$, $\tau \cong \tau^*$, and $\tau > \tau^*$, where τ^* is the delay time, and we match the three solutions sequentially to get a complete description of the delay and jump processes.

In the short time limit, the intensity is very small and (3.2) can be simply linearized around $I = 0$. In that case, the solution is

$$I(\tau) = I_0 \exp[v\tau^2 - 2\tau(1 - A)] \tag{3.3}$$

This solution remains exponentially small until τ approaches the critical time $\tau^* = 2(1 - A)/v$. In that domain, the solution (3.3) becomes

$$I(\tau) = I_0 \exp[v\tau(\tau - \tau^*)] \simeq I_0 \exp[2(1 - A)(\tau - \tau^*)] \tag{3.4}$$

When the critical time τ^* is reached, the solution of the linearized equation begins to diverge. The solution will jump from the vicinity of the trivial state to the vicinity of the nontrivial state. The slope of the jump diverges as $v \to 0$. Thus the evolution in this domain is dominated by the fast time τ. In the transition region, we introduce a new time $T = \tau - \tau^*$ and expand (3.2) around $v = 0$

$$dI/dT = 2I[1 - 2A + A/(1 + I)] \tag{3.5}$$

The solution of this equation can be written in implicit form as

$$T = \frac{1}{2(1 - A)}\left(\ln(I) + \frac{A}{1 - 2A}\ln[(1 - 2A)I - A + 1]\right) + c_1 \quad (3.6)$$

This solution must match the solution (3.4) as $T \to -\infty$. Thus we seek in the limit $T \to -\infty$ a solution that vanishes exponentially. From (3.6) we find

$$T \to \frac{1}{2(1 - A)}\ln(I) + c_1 \quad \text{as } I \to 0 \quad (3.7)$$

We invert this relation to express $I = I(T)$ and match this function with (3.4). This yields $c_1 = -\ln(I_0)/[2(1 - A)]$. The solution of (3.6) in the transition domain becomes

$$T = \frac{1}{2(1 - A)}\left(\ln(I/I_0) + \frac{A}{1 - 2A}\ln[(1 - 2A)I - A + 1]\right) \quad (3.8)$$

For the matching with the next domain, we need the asymptotic solution in the long time limit. Assuming that when $T \to +\infty$ it is the second term in (3.8) that diverges the most, we obtain

$$I \to \frac{1}{2A - 1}[1 - A + \exp(-T)] \quad (3.9)$$

After the jump has occurred, the time-dependent solution follows closely the upper branch of the static solution and we can solve (3.2) by a regular expansion in powers of the sweep rate

$$I(\tilde{\tau}, v) = I_0(\tilde{\tau}) + vI_1(\tilde{\tau}) + \ldots, \tilde{\tau} = v(\tau - \tau^*) \quad (3.10)$$

The time $\tilde{\tau}$ is the slow time that rules the evolution of the solution in this third domain. Inserting this expansion in (3.2) yields

$$I_0(\tilde{\tau}) = (1 - A + \tilde{\tau})/(2A - 1 - \tilde{\tau}) \quad (3.11)$$

This solution correctly matches (3.9) for $\tilde{\tau} \to 0$.

3.1.2 The critical case

It is clear that the previous section, and in particular the analysis in the transition layer, relies heavily on the assumption that $1 - A$ is an $\mathcal{O}(1)$ function. Let us consider now the case in which this assumption is invalid. If $A \simeq 1$, we can expect a different response of the system because $A = 1$ is a critical point. Of course, this is true only in the static regime. However, if the sweep rate is small and the distance, *in parameter space*, from the critical point is also small, we can consider that we are dealing with the perturbation of a critical point. Then we know from the discussion in Section 2.1 that there is critical slowing down and that the relevant time scale is not the physical time scale but a geometrical time scale that depends on the deviation from criticality. Thus we expect that the evolution will not occur on the fast time scale (associated with the time τ) but on a slower time scale. Since we do not know, a priori, which is the critical order of magnitude for A, let us assume a dependence

$$1 - A = v^\alpha a, \qquad 0 < v \ll 1, \quad a > 0 \qquad (3.12)$$

We rescale the intensity and the time in terms of the small parameter v:

$$s = v^\gamma \tau, \qquad I(\tau) = v^\beta J(s), \qquad I_0 = v^\beta J_0 \qquad (3.13)$$

where s is a new slow time. Inserting the scaling (3.12) and (3.13) into (3.2) yields, to dominant order, the equation

$$dJ/ds = 2J(-a + s - J) \qquad (3.14)$$

if we make the choice

$$\alpha = \beta = \gamma = 1/2 \qquad (3.15)$$

The solution of (3.14) is easily found by the substitution $J = 1/u^2$ that leads to a linear differential equation. The result is

$$J(s) = \frac{J_0 \exp[(s-a)^2]}{\exp(a^2) + 2J_0 \int_{-a}^{s-a} \exp(t^2)\, dt} \qquad (3.16)$$

This solution describes the whole evolution of the intensity and the smooth transition between the vicinity of the trivial solution and the vicinity of the nontrivial solution. Thus there is no longer a sharp jump transition between the two states. This is especially clear in Figure 3.1, where we plot the numerical solution of equation (3.2) versus $v\tau$ obtained in the two regimes and in the transition domain. The two extreme curves, for $A = 0.75$ and $A = 0.95$, are examples of the regular and critical cases, respectively. Another difference that is manifest in the two regimes is the scaling that affects the physical time in the transition region, be it sharp or smooth. In the regular case, the jump occurs on the fast time scale T that scales like v^{-1}, whereas in the critical case, the time s scales like $v^{-1/2}$. These two scalings are consistent with the general result $t^* = \mathcal{O}[|1 - A|/v]$ derived in the previous chapter. These dependences on v can be accessed experimentally. For instance, what can be measured with reasonable accuracy is the time at which the inflexion point occurs in the transition between the two solutions. Clearly, this time will display the two different scalings according to the initial value of A.

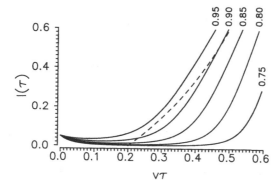

Figure 3.1 Equation (3.2) is integrated to obtain the intensity versus $v\tau$. The initial condition is $I(0) = 0.05$. The sweep rate is $v = 0.01$. Solid lines are labeled by A. The dotted line is the function $-1 + A/(1 - v\tau)$.

3.2 Weak external field

3.2.1 Constant amplitude

In the remaining sections of this chapter, we focus on the determination of the delay to the laser first threshold when various additional effects are taken into account. Furthermore, we assume that the sweep starts sufficiently far from the static critical point $A = 1$. Since we are concerned only with the determination of the delay, we may reduce the problem to the consideration of the linearized equation

$$dx/dt = x\mu(t), \qquad \mu(t) = \mu_0 + \varepsilon t,$$

$$0 < \varepsilon \ll 1, \quad \mu_0 < 0, \quad x(\mu_0) = x_0 \tag{3.17}$$

At this level, the swept gain or the swept loss problems are described by the same equation. The first modification we consider is the influence of a weak constant external field on the position of the laser first threshold. We denote by δ the amplitude of this weak field. The linearized equation becomes

$$dx/dt = \mu(t)x + \delta \tag{3.18}$$

The calculation will indicate which are the critical orders of magnitude of the external field amplitude. We define a new time $\tau = \mu(t)$ so that $\tau = 0$ is the time at which the static bifurcation is reached. In terms of the new time, (3.18) becomes $\varepsilon \, dx/d\tau = \tau x + \delta$ and its solution is

$$x(\tau) = x_0 \exp[(\tau^2 - \mu_0^2)/2\varepsilon] + (\delta/\varepsilon) \exp(\tau^2/2\varepsilon) \int_{\mu_0}^{\tau} \exp(-s^2/2\varepsilon) \, ds \quad (3.19)$$

In the limit $\varepsilon \to 0$, both τ and $|\mu_0|$ are much larger than $\sqrt{2\varepsilon}$. The dominant contribution to the integral appearing in the solution (3.19) is easily evaluated to be

$$\int_{\mu_0}^{\tau} \exp(-s^2/2\varepsilon) \, ds \sim \int_{-\infty}^{\infty} \exp(-s^2/2\varepsilon) \, ds = \sqrt{2\pi\varepsilon} \quad \text{as } \varepsilon \to 0 \quad (3.20)$$

The structure of (3.19) suggests that we define an exponentially small field amplitude

$$\delta = \varepsilon^p \exp(-k^2/2\varepsilon), \qquad k = \mathcal{O}(1), \quad p > 0 \quad (3.21)$$

With this definition and the asymptotic value of the integral, we write the solution (3.19) in the form

$$x(\tau) \sim x_0 \exp[(\tau^2 - \mu_0^2)/2\varepsilon] + \varepsilon^{p-1/2} \sqrt{2\pi} \exp[(\tau^2 - k^2)/2\varepsilon] \quad (3.22)$$

The properties of the delay will depend on which of the two exponentials diverges first.

- If $\mu_0^2 > k^2$, the second exponential diverges first. The delay is $k = \sqrt{2\varepsilon \ln(1/\delta)}$, which is $\mathcal{O}(1)$. It is reduced because it occurs for $\tau^* = k < -\mu_0$. Furthermore, the delay does not depend on the initial value of the control parameter.

- If $\mu_0^2 < k^2$, the first exponential diverges first and the delay occurs for $\tau^* = -\mu_0$. Thus, the delay is maximum and it depends on the initial value of the control parameter but not on δ.

We can summarize these conclusions as follows. There exists a critical field amplitude

$$\delta_c = \varepsilon^p \exp(-\mu_0^2/2\varepsilon), \qquad p > 1/2, \quad \mu_0 = \mathcal{O}(1) \quad (3.23)$$

such that if $\delta < \delta_c$, the delay is maximum, independent of the external field but dependent on the initial value of the control parameter; whereas if $\delta > \delta_c$, the delay is reduced, independent of the initial value of the parameter but dependent on the field amplitude.

The fact that the delay can be modified by an exponentially small external field could be bad news, as it implies that the delay is a rather sensitive (though not singular) property of the system. However, delays are now easily measured

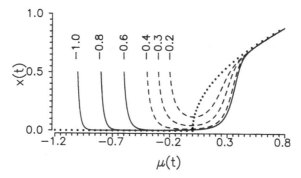

Figure 3.2 Solutions of equation (3.24) for $x(0) = 0.5$, $\varepsilon = 0.01$, and $\delta = 0.001$. The curves are labeled by μ_0. The dotted lines are the functions $x = 0$ and $x = \sqrt{\mu_0 + vt}$. The solid lines correspond to $\mu_0^2 > k^2$, the dashed lines correspond to $k^2 > \mu_0^2$.

and the dependence on even exponentially small terms makes them a sensitive way to test theoretical models and compare them with experimental results.

To illustrate some of the results obtained in this section, the following equation has been integrated numerically

$$dx/dt = x[\mu(t) - x^2] + \delta, \qquad \mu(t) = \mu_0 + \varepsilon t \qquad (3.24)$$

for a constant δ and different μ_0. The nonlinear term x^2 has been added to avoid an irrelevant divergence. The results are displayed in Figure 3.2, where the variation of the delay as a function of μ_0 is clearly seen.

3.2.2 Periodic amplitude

In general, when an external field is sent in a lasing cavity, the injected field and the lasing field do not have the same frequency. It requires very special conditions (and, usually, extreme ability in the laboratory) to match the two field frequencies. Thus our next level of sophistication is to consider an external field whose frequency differs from the cavity lasing field by ω. The linear problem becomes

$$dx/dt = x\mu(t) + \delta \cos(\omega t) \qquad (3.25)$$

We analyze the influence of the new parameter ω on the delay. In terms of the time $\tau = \mu(t)$, the solution of (3.25) is

$$x(\tau) = x_0 \exp[(\tau^2 - \mu_0^2)/2\varepsilon]$$

$$+ \varepsilon^p \exp[(\tau^2 - k^2)/2\varepsilon] \int_{\mu_0}^{\tau} \exp(-s^2/2\varepsilon) \cos[\omega(s - \mu_0)/\varepsilon] \, ds \quad (3.26)$$

where we have used the definition (3.21) of the exponentially small amplitude δ. The discussion of the delay requires a discussion of the integral in (3.26), which in turn depends on the ratio ω/ε. This discussion uses the property

$$\int_0^{+\infty} \exp(-b^2 s^2) \cos(ax) \, dx = (\sqrt{\pi}/2b) \exp[-(a/2b)^2] \quad (3.27)$$

First, if $\omega \ll \varepsilon$, the cosine function is approximated by 1 and the periodicity of the external field does not affect the delay.

Second, if $\omega = \mathcal{O}(\varepsilon)$, we introduce an $\mathcal{O}(1)$ frequency $\Omega = \omega/\varepsilon$. Then using $\cos[\Omega(s - \mu_0)] = \cos(\Omega\mu_0)\cos(\Omega s) - \sin(\Omega s)\sin(\Omega\mu_0)$, only the even function of s contributes asymptotically and we get from (3.27) the result

$$\int_{\mu_0}^{\tau} \exp(-s^2/2\varepsilon) \cos[\omega(s - \mu_0)/\varepsilon] \, ds \sim \cos(\Omega\mu_0) \int_{\mu_0}^{\tau} \exp(-s^2/2\varepsilon) \cos(\Omega s) \, ds$$

$$= \sqrt{2\pi\varepsilon} \cos(\Omega\mu_0) \quad \text{as } \varepsilon \to 0 \quad (3.28)$$

Hence the asymptotic solution (3.26) becomes

$$x(\tau) \sim x_0 \exp[(\tau^2 - \mu_0^2)/2\varepsilon] + \varepsilon^{p-1/2} \sqrt{2\pi} \exp[(\tau^2 - k^2)/2\varepsilon] \cos(\omega\mu_0/\varepsilon) \quad (3.29)$$

If $\mu_0^2 < k^2$, the situation is similar to the constant injected field and there is no effect of the periodicity of the injected field. However, if $\mu_0^2 > k^2$, the delay is reduced and independent of the initial value of the control parameter but it depends on the two parameters of the injected field. The delay depends on the field amplitude because it occurs for $\tau^* = k$. In addition, the divergence is toward $+\infty$ if $\cos(\omega\mu_0/\varepsilon) > 0$ and toward $-\infty$ if $\cos(\omega\mu_0/\varepsilon) < 0$. To understand this result, it is necessary to realize that the static nonlinear problem, which is $x' = x\mu - x^3$, has the trivial solution $x = 0$ for $\mu < 0$ and two more solutions $x = \pm \sqrt{\mu}$, making a pitchfork bifurcation, for $\mu > 0$. Because the delay occurs for $\mu > 0$, there are two possible branches available to the system. Both solutions correspond to the same intensity but the electric fields have a phase difference of π. Thus the sign of $\cos(\omega\mu_0/\varepsilon)$ selects the phase of the field.

Third, if $\omega = \mathcal{O}(\varepsilon^{1/2})$, another asymptotic behavior appears. Again, we introduce an $\mathcal{O}(1)$ frequency $\Omega = \omega/\sqrt{\varepsilon}$. In this case the integral in (3.26) has a different expansion:

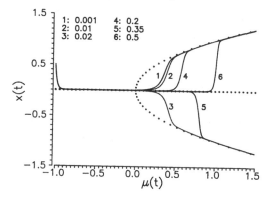

Figure 3.3 Solutions of equation (3.33) for $x(0) = 0.5$, $\varepsilon = 0.01$, $\mu_0 = -1$, and $\delta = 0.001$. The curves are labeled by the modulation frequency. The dotted lines are the functions $x = 0$ and $x = \pm\sqrt{\mu_0 + vt}$.

$$\int_{\mu_0}^{\tau} \exp(-s^2/2\varepsilon) \cos[\Omega(s - \mu_0)/\sqrt{\varepsilon}]\, ds$$

$$\sim \cos(\Omega\mu_0/\sqrt{\varepsilon}) \int_{-\infty}^{+\infty} \exp(-s^2/2\varepsilon)\cos(\Omega s/\sqrt{\varepsilon})\, ds$$

$$= \sqrt{2\pi\varepsilon}\exp(-\Omega^2/2)\cos(\Omega\mu_0/\sqrt{\varepsilon}) \quad \text{as } \varepsilon \to 0 \quad (3.30)$$

From this result it follows that the asymptotic solution of the linearized equation (3.25) in this regime of oscillation frequencies is

$$x(\tau) \sim x_0 \exp[(\tau^2 - \mu_0^2)/2\varepsilon]ds$$
$$+ \varepsilon^{p-1/2}\sqrt{2\pi}\exp[(\tau^2 - k^2 - \omega^2)/2\varepsilon]\cos(\omega\mu_0/\varepsilon)du$$
$$(3.31)$$

The difference with the previous case occurs if $\mu_0^2 > k^2 + \omega^2$. In that case the critical time that determines the delay is $\tau^* = \sqrt{k^2 + \omega^2}$. Hence the delay is now also controlled by the frequency of the injected field.

Fourth, if $\omega \geq \mathcal{O}(1)$, the effect of the injected field averages out and does not influence the delay any longer. In this limit, the integral in (3.26) is negligible and the solution takes the form

$$x(\tau) \sim x_0 \exp\left[(\tau^2 - \mu_0^2)/2\varepsilon\right] \quad (3.32)$$

with a field-independent delay.

The role of the modulation frequency ω is illustrated in Figure 3.3, where the equation

$$dx/dt = x[\mu(t) - x^2] + \delta\cos(\omega t), \qquad \mu(t) = \mu_0 + \varepsilon t \quad (3.33)$$

has been integrated for various frequencies, all other parameters being kept constant. Here again the nonlinear term x^3 has been subtracted to avoid irrelevant divergencies.

3.2.3 Stochastic amplitude

Until now, we have assumed that the injected field is a coherent field. Therefore it could be treated as a constant amplitude, modulated or not, depending on the mismatch between the frequencies of the injected and the cavity fields. However, the lasing field can also be perturbed by a random field that can be either generated inside the cavity by the atoms or injected from the outside. Although such a field is unavoidable in practical situations, it has to remain small if the laser is going to operate at all. Hence, the problem of the lasing first threshold in the presence of a stochastic perturbation can be formulated as

$$dx/dt = x\mu(t) + \xi(t), \qquad \mu(t) = \mu_0 + vt, \qquad \mu_0 < 0, \quad v > 0 \quad (3.34)$$

This is now a stochastic differential equation, or Langevin equation. We seek to determine the average of the stochastic variable $x^2(t)$ because it is proportional to the intensity, the quantity most easily measured in a laser experiment. Therefore we must characterize the noise source $\xi(t)$. The simplest choice is the white noise for which

$$\langle \xi(t) \rangle = 0, \qquad \langle \xi(t)\xi(t') \rangle = 2D\delta(t - t') \quad (3.35)$$

The first relation indicates that the stochastic source vanishes on average. The second relation implies the independence of the stochastic source at different times (this is the white noise property) and gives the second moment of the stochastic source. The choice (3.35) is only one type of noise source. Many other possibilities exist – usually motivated by physical considerations. A clear exposition of this subject can be found in the textbooks of Horsthemke and Lefever [2] and Gardiner [3], as well as in an exhaustive review of the field presented in the three volumes compiled by Moss and McClintock [4].

We analyze the Langevin equation (3.34) by the method of moments. Although this method is limited in its predictive capability, it has the advantage of elegance and simplicity. Furthermore, it is sufficient for a discussion of the position of the laser first threshold, which is the focus of this chapter. Another approach, involving the use of generating functions, has been developed by Torrent and San Miguel [5].

The problem formulated by (3.34) and (3.35) is not completely specified. We still have to discuss the initial condition. There are two possibilities: Either the stochastic source $\xi(t)$ can be switched on and off at will or it is always present. In this section we analyze the former case defined as follows. First, we fix the initial value of the control parameter $\mu = \mu_0$ and let the system evolve in the absence of the stochastic source $\xi(t)$ until a steady state is reached. That steady state will be the initial condition $x(0)$ for the swept problem. Second, the sweep begins with the initial condition $x(0)$ and the noise source is turned on.

It is quite easy to integrate (3.34) and (3.35) in this case because the formal solution of (3.34) is

$$x(t) = x(0)\exp[B(0, t)] + \int_0^t \xi(s)\exp[B(s, t)]\,ds$$

$$x^2(t) = x^2(0)\exp[2B(0, t)] + 2\exp[2B(0, t)]\int_0^t x(0)\xi(s)\exp[-B(0, s)]\,ds$$

$$+ 2\int_0^t ds \int_0^s \xi(s)\xi(u)\exp[2B(s, t)]\exp[B(u, s)]\,du \qquad (3.36)$$

where $B(s, t) = \int_s^t \mu(s)\,ds$. Using the autocorrelation property (3.35), we find that $\langle x(0)\xi(t)\rangle$ vanishes and

$$2\int_0^t \int_0^s \langle\xi(s)\xi(u)\rangle\exp[2B(s, t)]\exp[B(u, s)]\,ds\,du$$

$$= D\sqrt{\pi/v}\exp(\zeta^2)[\mathrm{Erf}(\zeta) - \mathrm{Erf}(a)] \qquad (3.37)$$

where we have used the definitions

$$\zeta(t) = \mu(t)/\sqrt{v}, \qquad a = \zeta(0) < 0, \qquad \mathrm{Erf}(\zeta) = (2/\sqrt{\pi})\int_0^\zeta \exp(-y^2)\,dy \qquad (3.38)$$

It is now easy to obtain an expression for the average intensity

$$\langle x^2(t)\rangle = x^2(0)\exp(\zeta^2 - a^2) + D\sqrt{\pi/v}\exp(\zeta^2)[\mathrm{Erf}(\zeta) - \mathrm{Erf}(a)] \quad (3.39)$$

The delay is determined by the condition $\langle x^2(t^*)\rangle = x^2(0)$. We could use the more general definition $\langle x^2(t^*)\rangle = I_{th}$, where I_{th} is some predefined threshold intensity corresponding, for instance, to the detection threshold of the device used to measure the intensity. However, there is no gain in the understanding of the phenomena when $I_{th} \neq x^2(0)$. With the simplest choice, the equation that determines the delay is $\langle x^2(t^*)\rangle = x^2(0)$, which implies

$$\exp(-z^2) - b\,\mathrm{Erf}(z) = \exp(-a^2) - b\,\mathrm{Erf}(a) \qquad (3.40)$$

where $z = \zeta(t^*)$ and $b = [D/x^2(0)]\sqrt{\pi/v}$.

Let us consider the two limits of large and small noise amplitude. If $b \geq 1$ (large noise limit), the only solution is $z \cong -a$ and there is no delay: A blurred transition takes place at the static bifurcation point. In the small noise limit,

$b \ll 1$, another major difference occurs. In the deterministic case, the delay $\mu(t^*)$ increases linearly with the decrease of the initial condition $\mu(0)$. In the noisy case, there is a saturation effect. Indeed, in the double limit $v \to 0$ and $b \to 0$, we have

$$\exp(-z^2) \cong b[1 + \mathrm{Erf}(z)] \cong b[2 + \mathcal{O}(e^{-z})] \tag{3.41}$$

To obtain this result, we used the fact that $\mathrm{Erf}(a) \to -1$ because $a \to -\infty$ as $v \to 0$. Hence there is a maximum delay given by $z_{max} \simeq \sqrt{\ln(1/2b)}$. In terms of the physical parameters, this relation becomes

$$\mu(t^*)_{max} = \sqrt{v \ln\left(\frac{x^2(0)\,\sqrt{v/\pi}}{2D}\right)} \tag{3.42}$$

An interesting experimental study of this problem can be found in [6], where an analog simulation was set up to analyze the phenomena associated with noise and swept parameter in a controlled way. An extension of this analysis can be found in Reference [7].

References
[1]   W. Scharpf, M. Squicciarini, D. Bromley, C. Green, J. R. Tredicce, and L. M. Narducci, *Opt. Commun.* **63** (1987) 344.
[2]   W. Horsthemke and R. Lefever, *Noise-Induced Transitions* (Springer, Heidelberg, 1984).
[3]   C. W. Gardiner, *Handbook of Stochastic Methods* (Springer, Heidelberg, 1983).
[4]   F. Moss and P. V. E. McClintock, eds., *Noise in Nonlinear Dynamical Systems* (Cambridge University Press, Cambridge, 1989).
[5]   M. C. Torrent and M. San Miguel, *Phys. Rev. A* **38** (1988) 245.
[6]   R. Mannella, F. Moss, and P. V. E. McClintock, *Phys. Rev. A* **35** (1987) 2560.
[7]   H. Zeghlache, P. Mandel, and C. Van den Broeck, *Phys. Rev. A* **40** (1989) 286.

4

Optical bistability: Constant input

4.1 Introduction

Usage has reserved the expression *optical bistability* mostly for coherently driven passive systems. An atomic system is called *passive* when there is no population inversion. This is the case, for example, of a system at thermal equilibrium. On the contrary, the laser that we have described in Chapter 1 is a driven active system, because a population inversion is created. The laser equations we have studied so far describe an incoherently driven laser. However, nothing prevents driving a lasing cavity with a coherent field emitted by another laser. This is realized in a whole class of lasers that are used mainly as frequency converters. In optical bistability (OB), the driving field is usually a coherent field. Thus we have to account for two differences between the laser and the optically bistable system: (1) In OB, there is no inversion of population: $\langle |A^2| \rangle > \langle |B^2| \rangle$, in the notation of Chapter 1. Hence, in the absence of interaction with a field, the population difference $\mathcal{D}$ relaxes toward a negative value. (2) The pumping is coherent, meaning that an external laser field is added to the cavity field.

Using the single-mode equations (1.48)–(1.50), we have to change the sign of $\mathcal{D}$ and $\mathcal{D}_a$ (which amounts to keeping D defined in (1.34) as it is but changing P into $-P$ and A into $-A$) and to add a source term in the equation for the complex field amplitude. When this is done, the reduced equations are

$$dE/dt \equiv E' = -\kappa[(1 + i\Delta)E + 2CP - E_i] \tag{4.1}$$

$$P' = -(1 + i\delta)P + ED \tag{4.2}$$

$$D' = \gamma[1 - D - (1/2)(E^*P + EP^*)] \tag{4.3}$$

where we have written $2C$ instead of $-A$ as is usual in OB. To simplify the notation, we use t for the time that was denoted by τ in Chapter 1. E_i is the fraction of the driving field entering the cavity. If the in-coupling mirror has a

37

field transmission coefficient $T_f < 1$, the injected field inside the cavity E_i is related to the injected field outside the cavity E_o by the relation $E_i = E_o \sqrt{T_f}$. The amplitude E_i of the injected field is real, giving a reference to the phases of the cavity field and of the atomic polarization. It also means that the reference frequency chosen in this formulation is the injected field frequency ω_i so that we have the definition

$$\Delta = (\omega_c - \omega_i)/\gamma_c, \qquad \delta = (\omega_a - \omega_i)/\gamma_\perp \qquad (4.4)$$

Two cases have to be considered when dealing with OB: the absorptive limit, in which $\Delta = \delta = 0$, and the dispersive case, in which at least one of the detuning functions does not vanish. As will be seen later in this book, many limits can be considered in the dispersive case. However, for most of this chapter, it is sufficient to consider the absorptive limit

$$E' = -\kappa(E + 2CP - E_i) \qquad (4.5)$$

$$P' = -P + ED \qquad (4.6)$$

$$D' = \gamma(1 - D - EP) \qquad (4.7)$$

4.2 Steady states

Much of what will be derived in the following sections is in fact independent of the explicit form of the equations and depends only on geometrical properties of the steady solutions. Thus, the equations (4.5)–(4.7) should be considered merely as a convenient support for the derivation of generic equations. We will show later that other equations are amenable to similar treatments. The essential feature of (4.5)–(4.7) appears already in their steady solution

$$D = 1/(1 + E^2) \qquad (4.8)$$

$$P = E/(1 + E^2) \qquad (4.9)$$

$$E_i = E + 2CE/(1 + E^2) \qquad (4.10)$$

The steady cavity field amplitude E is the solution of a cubic equation. These solutions are displayed in Figure 4.1.

There is a critical C, which we write C_{crit}, separating two domains. In the domain $C \leq C_{crit}$, the solution E versus E_i is monostable with a slope at the inflexion point that increases as C approaches C_{crit}. In other terms, the cubic equation for E has only one real root. However, for $C > C_{crit}$, there is a domain in which all three roots are real. This domain is called the *bistable domain* because in many examples the upper and the lower branches are stable while the intermediate branch is unstable. The reader should be warned against a common belief that is wrong. For 1-D systems, if the steady state solution of the

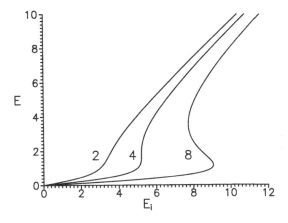

Figure 4.1 Steady state solutions from equation (4.10). The curves are labeled by the value of C.

evolution equation is a cubic, the intermediate branch is unstable while the upper and lower branches are stable. Otherwise, the solution to the *perpetuum mobile* would have been found. However, in more than one dimension, this conclusion is not valid in general. Examples (that are not pathological in any sense) have been found in which the intermediate branch is stable. This is because in N-dimensional ($N > 1$) systems, what appears as the intermediate branch in the (E, E_i) plane is only the projection of an N-dimensional surface. Similarly, there are many known examples in which part of the lower and/or upper branches are unstable. In some instances, the upper branch can even be entirely unstable.

To determine the coordinates of the two limit points for $C > C_{crit}$, we compute the derivative of E versus E_i

$$dE/dE_i = (1 + E^2)/(3E^2 - 2EE_i + 1 + 2C) \qquad (4.11)$$

Limit points are characterized by an infinite derivative that occurs for

$$E_{\pm} = (1/3)\left(E_i \pm \sqrt{E_i^2 - 3(1 + 2C)}\right) \qquad (4.12)$$

Hence the boundary of existence for bistability is $E_+ = E_- \equiv E_{crit}$; that is, $3E_{crit} = E_{i,crit}$ and $E_{i,crit}^2 = 3(1 + 2C_{crit})$. Combining this relation with the steady state equation (4.10) yields $C_{crit} = 4$. There is a unique point of vertical slope if

$$C_{crit} = 4, \qquad E_{crit} = \sqrt{3}, \qquad E_{i,crit} = 3\sqrt{3} \qquad (4.13)$$

For $C > C_{crit}$, we solve the steady state equation (4.10) and the relation (4.12) for E and obtain the equation

$$E_{\pm}^4 + 2(1 - C)E_{\pm}^2 + 1 + 2C = 0 \qquad (4.14)$$

from which the cavity field coordinates at the two limit points are obtained

$$E_+^2 = C - 1 + \sqrt{C(C - 4)}, \qquad E_-^2 = \frac{1 + 2C}{C - 1 + \sqrt{C(C - 4)}} \qquad (4.15)$$

Although it is possible to investigate the linear stability of the steady state satisfying (4.10), the fact that we cannot write even the steady solutions explicitly in reasonably simple form implies that the stability analysis would be a vain and formal exercise at this point. Stability analyses will be performed in the limits studied in the next sections.

4.3 Reduction to one dimension

In this section we consider two limits in which a 1-D reduction of the 3-D equations (4.5)–(4.7) is possible. The simplest limits are those in which one of the decay rates is much smaller than the other two. For instance, in the good cavity limit, it is assumed that $\kappa \ll 1$ and $\kappa \ll \gamma$. This leads to a closed equation for the cavity field E. Another important limit, especially in OB, is the bad cavity limit in which $\gamma \ll 1$ and $\gamma \ll \kappa$. It leads to a closed equation for the population difference D. These are classical examples in which two out of the three equations can be adiabatically eliminated in a fairly simple way. Other limits are also amenable to an asymptotic treatment and can lead to a 1-D equation without strong inequalities between the decay rates. They are studied in Sections 4.5 and 5.3.

4.3.1 Good cavity limit

By good cavity, we mean a cavity that is nearly perfect, that is, a cavity whose mirror reflectivities are close to unity. For such a cavity, the field damping rate is small compared to the atomic damping rates. Hence we characterize such a cavity by the inequalities

$$\kappa \ll 1, \qquad \kappa \ll \gamma, \qquad \gamma = \mathcal{O}(1) \qquad (4.16)$$

Only the first two inequalities are necessary to define the good cavity limit. The third inequality is necessary to complete the definition of the limit we use, but there are other choices for the magnitude of γ. For instance $\gamma \ll 1$ and $\gamma \gg 1$ can be compatible with $\kappa \ll 1$ and $\kappa \ll \gamma$.

We seek solutions of the three evolution equations (4.5)–(4.7) in the form of power series $Z(\kappa, t) = Z(\kappa, \tau) = Z_0(\tau) + \kappa Z_1(\tau) + \mathcal{O}(\kappa^2)$, where τ is the slow time defined by $\tau = \kappa t$ and Z is any of the three dynamical variables. Inserting this expansion into (4.5)–(4.7) leads, to dominant order, to a single equation

$$dE_0/d\tau = -E_0 - 2CE_0/(1 + E_0^2) + E_i \qquad (4.17)$$

4.3.2 Bad cavity limit

Another useful limit is the bad cavity limit in which

$$\gamma \ll 1, \qquad \gamma \ll \kappa, \qquad \kappa = \mathcal{O}(1) \qquad (4.18)$$

Here also only the first two inequalities are necessary to define the bad cavity limit. The third inequality completes the definition of the limit that we use. In laser physics, the interest in the bad cavity limit is seen by many physicists as somewhat perverse because it is the domain in which instabilities occur. These instabilities lead to a periodic or chaotic lasing intensity that is usually not the kind of property one looks for when building a laser. In OB, on the contrary, the constraints are different and the emphasis is on dealing with cheap devices. The typical example is a sample of semiconductor with cleaved uncoated end faces. In that case, the difference between the refractive index of the sample and that of the air will serve as a (poor) mirror. Thus the active variable will be the population difference D, which in the case of a semiconductor device, is related to the nonlinear refractive index. Bad cavity devices usually operate in the dispersive limit. Hence we derive the relevant 1-D equation using the complete equations (4.1)–(4.3). We seek solutions in the form $Z(\gamma, t) = Z(\gamma, \tau) = Z_0(\tau) + \gamma Z_1(\tau) + \mathcal{O}(\gamma^2)$, where τ is the slow time defined by $\tau = \gamma t$ and Z is any of the three dynamical variables. This leads to

$$dD_0/d\tau = 1 - D_0 - D_0|E_0|^2/(1 + \delta^2) \qquad (4.19)$$

$$|E_0|^2 \left[\left(1 + \frac{2CD_0}{1 + \delta^2}\right)^2 + \left(\Delta - \frac{2C\delta D_0}{1 + \delta^2}\right)^2 \right] = E_i^2 \qquad (4.20)$$

In the expression (4.20) that relates the cavity field to the input field, the first bracket, $1 + 2CD_0/(1 + \delta^2)$, accounts for the absorptive properties of the medium amd the second bracket, $\Delta - 2C\delta D_0/(1 + \delta^2)$, accounts for the dispersive properties. We introduce the decomposition $D_0 = 1 - D_e$ with $|D_e| \ll 1$, and we take the dispersive limit defined by

$$\delta^2 \gg 1, \qquad \delta^2 \gg \Delta^2, \qquad 2C = \mathcal{O}(\Delta\delta) \qquad (4.21)$$

In this limit, we obtain the approximate equations

$$dD_e/d\tau \cong -D_e + |E_0|^2/\delta^2 \qquad (4.22)$$

$$|E_0|^2 \left[\delta^2 + \left(\delta\Delta - 2C + 2CD_e \right)^2 \right] = E_i^2 \delta^2 \qquad (4.23)$$

where we have linearized the evolution equation (4.22) with respect to D_e. After the rescaling

$$n = 2CD_e/\delta, \qquad \theta = -\Delta + 2C/\delta, \qquad I = 2CE_i^2/\delta^3 \qquad (4.24)$$

equation (4.22) can eventually be written in the simple form

$$n' = -n + I/[1 + (n - \theta)^2] \qquad (4.25)$$

4.4 The osculating parabola

One obvious property of the two approximate 1-D equations we have derived in the previous section is that they preserve the steady state hysteresis that characterizes OB. In many applications of OB, the main questions are related to the switching dynamics of the devices. More generally, the behavior of the device in the vicinity of the limit points is the main concern because it will deeply affect the usability of the device. As will be seen in this section, the vicinity of the limit point is characterized by critical slowing down. We have already seen in Chapters 2 and 3 that critical slowing down may deeply affect the dynamical response of the system to a time variation of the control parameter. The same occurs near the limit points.

There are many systems, optical or not, that are bistable. This property merely reflects the occurrence of a certain class of nonlinear interactions among the constituents of the system that displays bistability. Therefore, it is useful to find a description of the bistable system that does not depend on its physical mechanisms. This is exactly the line of thought we followed in Chapter 3 to analyze a set of generic problems. The same is possible in OB. The procedure to be used is quite simple to explain, even though it may involve some heavy algebra. Let us consider, for example, the good cavity limit described by (4.17).

1. We define a small parameter ε as the deviation of E_0 from its value at the limit point. We also expand the injected field in powers of ε.

2. We insert these expansions in (4.17) and retain contributions up to the second order.

3. We rescale the field variables in such a way that the coordinates of the limit point are $(1,1)$. Let x be the rescaled field and μ the control parameter proportional to the injected field.

4. We rescale time by requiring that the coefficient of x^2 is unity.

After these steps, the evolution equation in the vicinity of the lower limit point (where the up-switching process begins) is

$$dx/dt = x^2 - 2x + \mu \qquad (4.26)$$

Since this description is local, the upper branch of the complete OB response curve has receded to $+\infty$. Indeed, the rescaling that maps the vicinity of $\mathcal{O}(\varepsilon)$ near the limit point into an $\mathcal{O}(1)$ domain maps the previously $\mathcal{O}(1)$ functions into $\mathcal{O}(1/\varepsilon)$ functions. To dominant order, the upper branch is at infinity, a useful property that will be exploited.

If we follow the same line of reasoning for the vicinity of the upper limit point, we obtain in the parabolic approximation

$$dx/dt = -x^2 + 2\mu x - 1 \qquad (4.27)$$

In this case, it is the lower branch of the complete OB response curve that has been removed from the description, being $\mathcal{O}(\varepsilon)$.

The steady state solution of either (4.26) or (4.27) is a parabola that is tangent to the exact steady state solution of (4.10) at the limit point. Such a parabola is known in analytical mechanics as the *osculating parabola*.

4.4.1 Steady state solutions and nonlinear stability

We concentrate our analysis on equation (4.26), but qualitatively all the results we obtain remain true for equation (4.27), which describes the down-switching process.

The steady state solutions of (4.26) are

$$x_\pm = 1 \pm \sqrt{1 - \mu} \qquad (4.28)$$

A nonlinear equation of the type (4.26) can be solved analytically in a number of cases. Indeed, the change of variables $x = -u'/u$ yields for the variable u the linear second-order equation

$$u'' + 2u' + \mu u = 0 \qquad (4.29)$$

Thus we are able to produce an analytic solution of (4.26) for all $\mu = \mu(t)$ for which (4.29) has a known analytic solution. We begin with the simplest case, namely a constant value of the control parameter. Let $\mu < 1$ and $x(0) = x_+ + \beta$ where the deviation β can be either positive or negative. The solution of (4.26) is

$$x(t) = x_- + \frac{(x_+ - x_-)(x_+ - x_- + \beta)}{x_+ - x_- + \beta(1 - e^{2t\sqrt{1-\mu}})} \qquad (4.30)$$

Because we have the exact solution, we perform a complete stability analysis, that is, a nonlinear stability analysis. Four cases arise with $\mu < 1$.

First, if $\beta < 0$, the long time limit of (4.30) is $x_- + \mathcal{O}(e^{-2t\sqrt{1-\mu}})$. Thus for an initial condition $x(0)$ below the upper branch of the parabola, the lower branch x_- is a stable attractor.

Second, if $\beta = 0$, the system remains on its initial condition forever.

Third, if $\beta > 0$, a most interesting phenomenon occurs. The solution $x(t)$ diverges in a finite time T given by

$$T = \frac{1}{2\sqrt{1-\mu}}\left(\ln(x_+ - x_- + \beta) - \ln(\beta)\right) \qquad (4.31)$$

Thus, the branch of solutions x_+ is unstable against arbitrarily small perturbations. The divergence in a finite time results from the nonlinearities. Note that the upper branch in this local description is the intermediate branch of the full bistable curve. The solution (4.30) and the critical time (4.31) illustrate very clearly the concept of critical slowing down in the vicinity of a limit point. Indeed, from (4.30) it appears that $2\sqrt{1-\mu}$ is the inverse of the relaxation time. If $\mu \to 1$ with β finite, the relaxation time diverges and so does the critical time T. The point $(x, \mu) = (1, 1)$ is a critical point in the sense that two solutions coincide at that point. It is not a steady bifurcation point as defined in the beginning of Chapter 2 because there is no real solution on one side of the critical point. Nevertheless, it suffices that a point be a critical point to induce critical slowing down.

Fourth, if $|\beta| \ll 1$ and $1 - \mu = \mathcal{O}(1)$, another singularity appears since the divergence time (for $\beta > 0$) and the relaxation time (for $\beta < 0$) are of the form

$$\text{Characteristic time} = \frac{\ln\left(1/|\beta|\right)}{2\sqrt{1-\mu}} + \mathcal{O}(1) \qquad (4.32)$$

To distinguish this type of divergence from the usual critical slowing down, the expression *noncritical slowing down* has been coined. The logarithmic divergence is associated with the vicinity of the unstable solution x_+. Here, the change of stability is not associated with a particular value of the control parameter but with the initial condition: the line x_+ is a separatrix, being the boundary between the basins of attraction of the solutions x_- and $+\infty$ (which stands for the upper branch in this model). Thus both divergences are related to a degeneracy. Critical slowing down occurs if the control parameter approaches a critical point where two solutions coincide, whereas noncritical slowing down results from the degeneracy of two basins of attraction along a separatrix.

4.4.2 Long pulses

In the eighties, hopes were high that OB would pave the way to the optical computer or, more modestly, to optical signal processing. This hope has motivated much research and a large number of publications to which it is impossible to do full justice. References [1] to [4] are reviews of that literature. Although there is still active research in this direction, we have learned that even if success is at the end of the road, the road will be much longer than expected. The potential of application for OB has oriented its development in a direction that we begin to explore. Many problems are linked with the questions of how fast can a bistable device switch or of how stably can it be maintained on either the lower or the upper branch.

The simplest problem we consider is the response of the bistable system to a long switching pulse. In other words, we imagine that the system has been prepared in its stable state, which we take as the initial condition. Then we suddenly increase the control parameter μ (which is proportional to the injected field amplitude) to a value above the limit point. The question is how long does it take for the system to reach the upper branch of the complete bistable curve, shifted at $+\infty$ in this local description. Thus we have to solve (4.26) with the conditions

$$
\begin{aligned}
t = 0 &: \mu = \mu_0 < 1, \quad x(0) = x_-(\mu_0) \\
t > 0 &: \mu = \mu_1 > 1
\end{aligned}
\tag{4.33}
$$

This is easily done and the result is

$$
x(t) = \frac{(1 - i\Omega)(\alpha + i\Omega) - (1 + i\Omega)(\alpha - i\Omega)e^{-2i\Omega t}}{\alpha + i\Omega - (\alpha - i\Omega)e^{-2i\Omega t}}
\tag{4.34}
$$

where we have introduced the definitions

$$
\alpha = \sqrt{1 - \mu_0}, \qquad \Omega = \sqrt{\mu_1 - 1}
\tag{4.35}
$$

The solution (4.34) diverges at a critical time T such that $\alpha + i\Omega = (\alpha - i\Omega)e^{-2i\Omega T}$, or equivalently,

$$
\cos(2\Omega T) = \frac{\alpha^2 - \Omega^2}{\alpha^2 + \Omega^2}, \qquad \sin(2\Omega T) = -\frac{2\alpha\Omega}{\alpha^2 + \Omega^2}
\tag{4.36}
$$

We analyze these equations in three limit cases.

First, $\alpha = 0$, $\Omega \neq 0$; then we have $T = \pi/2\Omega \equiv T^*$. The critical value T^* is the jump duration since it is the time required to switch right from the limit point to $+\infty$, the upper branch of the complete bistable curve. This critical time is essentially model-dependent, and it may be either a good or a poor approximation. Therefore, we seek to derive quantities that are independent of T^*.

Second, $\alpha \ll 1, \Omega = \mathcal{O}(1)$; the physical situation corresponding to this limit is a bistable system held close to the limit point and a pulse of $\mathcal{O}(1)$ amplitude is added to the holding beam to induce the jump. In this limit, the solution of (4.36) is $T = T^* + \alpha/\Omega^2 + \mathcal{O}(\alpha^3)$. We introduce the destabilization time, $\Delta T \equiv T - T^*$, which describes the local part of the dynamics. The consideration of this destabilization time makes sense only if it captures the essentials of the dynamical evolution. Numerous comparisons with numerical models and with experiments indicate that ΔT is indeed a useful concept. Thus our first result is

$$\alpha \ll 1, \qquad \Omega = \mathcal{O}(1), \qquad \Delta T = T - T^* = \alpha/\Omega^2 + \mathcal{O}(\alpha^3) \quad (4.37)$$

Third, $\Omega \ll 1, \alpha = \mathcal{O}(1)$; these conditions describe a system held well below the limit point. The pulse added to the holding beam brings the control parameter barely above the critical point. In this limit, the solution of (4.36) is $T = \pi/\Omega + \mathcal{O}(1)$. Hence the destabilization time does not cancel exactly the minimum jump time T^*, and we have the solution

$$\Omega \ll 1, \qquad \alpha = \mathcal{O}(1), \qquad \Delta T = T - T^* = \pi/\Omega + \mathcal{O}(1) \quad (4.38)$$

This result was to be expected. Indeed, the condition $\Omega \ll 1$ indicates that the dynamics takes place close to the limit point. The dependence of the divergence and the destabilization times on $1/\Omega$ is a manifestation of the usual critical slowing down.

Two remarks should be made at this point. First, we note that both limits have produced a destabilization time inversely proportional to Ω though with different exponents. This has been confirmed experimentally, which also supports the usefulness of this approach. Second, the limit $\alpha \ll 1, \Omega = \mathcal{O}(1)$ has a further implication. If we consider the area of the destabilizing pulse $N = (\mu_1 - \mu_0)\Delta T$, using (4.37) we find

$$\alpha \ll 1, \qquad \Omega = \mathcal{O}(1), \qquad N = \alpha + \mathcal{O}(\alpha^3) = \sqrt{1 - \mu_0} + \mathcal{O}(\alpha^3)$$
$$(4.39)$$

The important aspect of this result is that the pulse area N does not depend on μ_1 and therefore it is independent of either the pulse length or amplitude. In optics, this means that what matters for destabilizing the system is only the total number of photons delivered to the system.

The first report on the pulse area law in optical bistability can be found in Reference [5], where it was obtained numerically for gaussian pulses with and without added noise. Experimental results along the lines of the theory presented here have been reported in [6] and [7]. The result (4.38) was derived initially for OB in [8], and an experimental verification has been published in [9]. The variation of the switching time scaling as given by the complete equation (4.36), including the two asymptotic results (4.37) and (4.38), has been studied numerically and experimentally in [10].

4.4.3 Short pulses

We consider now the response to a pulse of finite duration. The definition of the problem is

$$
\begin{aligned}
t = 0: \quad & \mu = \mu_0 < 1, \quad x = x_-(\mu_0) \\
0 < t < \tau: \quad & \mu = \mu_1 > 1 \\
t > \tau: \quad & \mu = \mu_0
\end{aligned}
\tag{4.40}
$$

During the action of the pulse ($0 < t < \tau$), the solution of (4.26) is

$$
x_1(t) = \frac{(1 + i\Omega)(i\Omega - 1 + x_-) + (1 - i\Omega)(i\Omega + 1 - x_-)e^{2i\Omega t}}{i\Omega - 1 + x_- + (i\Omega + 1 - x_-)e^{2i\Omega t}}
\tag{4.41}
$$

This solution reduces to $x_- \equiv x_-(\mu_0)$ for $t = 0$. After the pulse has been applied ($t > \tau$), the solution of (4.26) becomes

$$
x_2(t) = \frac{x_+(X - x_-) - x_-(X - x_+)e^{(x_+ - x_-)(t - \tau)}}{X - x_- - (X - x_+)e^{(x_+ - x_-)(t - \tau)}}
\tag{4.42}
$$

where we have used the definitions

$$
X \equiv x_1(\tau), \qquad x_\pm = 1 \pm \sqrt{1 - \mu_0}, \qquad \Omega = \sqrt{\mu_1 - 1} \tag{4.43}
$$

There are two obvious limits. If the pulse duration is very short, there is no switching and the system relaxes back to the stable lower branch. On the other hand, if the pulse duration is sufficiently long, switching necessarily occurs. Thus a critical pulse duration, $\tau = \tau^*$, separates the two domains. At criticality, the system neither switches up nor relaxes: The remaining possibility is that it stays where it has been brought by the pulse. This can only be the unstable branch x_+. This branch is unstable against perturbations but if the system is brought right on it, it stays there forever (in a deterministic description). Thus, the critical pulse duration is defined by the implicit equation $x^* \equiv x_1(\tau^*) = x_+$. For pulse durations longer than the critical pulse duration, the solution (4.42) diverges at the time $t_\uparrow$ defined by the condition

$$
X - x_- - (X - x_+)e^{(x_+ - x_-)(t_\uparrow - \tau)} = 0 \tag{4.44}
$$

Solving for $t_\uparrow$ gives the relation

$$
t_\uparrow = \tau + \frac{1}{x_+ - x_-} \ln \frac{X - x_-}{X - x_+} \tag{4.45}
$$

If the pulse duration exceeds only slightly the critical value, we find

$$
t_\uparrow = \frac{1}{x_+ - x_-} \ln(1/\varepsilon) + \mathcal{O}(1), \quad X = x_+ + \varepsilon, \quad 0 < \varepsilon \ll 1 \tag{4.46}
$$

Similarly, if the pulse duration is smaller than the critical value, the solution
(4.42) describes the relaxation of the system toward the stable state x_-

$$x_2(t) = \frac{x_- + x_+ e^{-\varphi(t)}}{1 + e^{-\varphi(t)}}, \qquad \varphi(t) = (x_+ - x_-)(t - \tau) - \ln \frac{X - x_-}{x_+ - X}$$

$$(4.47)$$

The relaxation time, $t_\downarrow$, is defined as usual by the condition $\varphi(t_\downarrow) = 1$ which
yields

$$t_\downarrow = \tau + \frac{1}{x_+ - x_-} \left(1 + \ln \frac{X - x_-}{x_+ - X} \right) \qquad (4.48)$$

If the pulse duration is only slightly less than the critical duration, the relaxation
time becomes

$$t_\downarrow = \frac{1}{x_+ - x_-} \ln(1/\varepsilon) + \mathcal{O}(1), \quad X \equiv x_+ - \varepsilon, \quad 0 < \varepsilon \ll 1 \quad (4.49)$$

Comparing (4.46) and (4.49) indicates that in the vicinity of the critical pulse
length τ^* both characteristic times coincide. If $\mu_1 - \mu_0 \leq \mathcal{O}(\varepsilon)$, the dominant
divergence of both characteristic times comes from the term $1/(x_- - x_+)$. It
is a power divergence typical of the vicinity of a critical point. However, if
$\mu_1 - \mu_0 = \mathcal{O}(1)$, the divergence of the characteristic times is only logarithmic
and this is a signature of noncritical slowing down. This property is displayed in
Figure 4.2, where equation (4.26) is integrated in the short pulse domain close
to the time τ^*. An experimental study of these concepts can be found in [10],
[11], and [12].

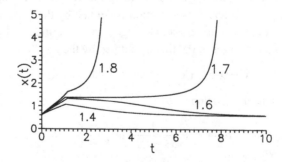

Figure 4.2 Noncritical slowing down. Equation
(4.26) is integrated in the short pulse domain (4.40)
with $\mu_0 = 0.85$ and a pulse duration $\tau = 1$. The
curves are labeled by the pulse amplitude μ_1.

4.5 Fully developed bistability

In Section 4.3 we have seen that if one decay rate is much larger than the other two [one decay rate is unity with the time scaling defined in (1.35)] it is easy to reduce the problem to a 1-D equation. This is because an inequality like $\kappa \ll \gamma$ implies the existence of two well separated time scales and the relevant dynamics occurs on the slowest time scale. The question that arises naturally is whether a strong inequality between the decay rates is necessary or only sufficient to ensure the reduction to a 1-D description. The answer is that the inequality is only sufficient, but by no means necessary [13]. To demonstrate this point, we analyze a simple case in which all three decay rates are comparable in magnitude. Let us consider absorptive bistability in the limit of a large hysteresis domain ($C \gg 4$) in a cavity that is neither good nor bad.

$$1/C = \varepsilon \ll 1, \quad \kappa = \mathcal{O}(1), \quad \gamma = \mathcal{O}(1) \tag{4.50}$$

We continue to focus our analysis on the limit point of the lower branch in the (E, E_i) plane. From (4.15) we find that the field near the lower limit point is $E_- = 1 + \mathcal{O}(\varepsilon)$. From the steady state equation (4.10), it then follows that the injected field near the lower limit point is $E_i = C + \mathcal{O}(1)$. This suggests that we seek solutions of the equations for optical bistability (4.5)–(4.7) with the following scaling

$$E = \mathcal{O}(1), \quad P = \mathcal{O}(1), \quad D = \mathcal{O}(1), \quad E_i = Y/\varepsilon = \mathcal{O}(1/\varepsilon) \tag{4.51}$$

We can set $\kappa = \gamma = 1$ without losing qualitative features of this problem. To dominant order in C, the evolution equations (4.5)–(4.7) become

$$\varepsilon E' = -\varepsilon E - 2P + Y \tag{4.52}$$

$$P' = -P + ED \tag{4.53}$$

$$D' = 1 - D - EP \tag{4.54}$$

The structure of (4.52) is already an indication that we are not dealing with a regular problem. The limit $\varepsilon \to 0$ is singular because the small parameter affects the highest order derivative in (4.52), which is E'. If $\varepsilon = 0$, equations (4.52)–(4.54) reduce to a single differential equation for D and the relations $2P = Y, E = Y/2D$. If we perform the linear stability analysis of the steady state solution, there are three roots if $\varepsilon \neq 0$ and only one root if $\varepsilon = 0$. However, the real parts of these roots are the decay rates of the system and the disappearance of two of them signifies that in the limit $\varepsilon \to 0$ there are diverging characteristic times. This is an alternative way in which an inequality between time scales can be induced. Let us see this mechanism in detail. We use (4.52) to eliminate P among the other two equations. This leads to

$$\varepsilon(E'' + 2E') + E(\varepsilon + 2D) - Y = 0 \qquad (4.55)$$

$$D' = 1 - D - EY/2 + \varepsilon E(E + E')/2 \qquad (4.56)$$

The steady state solutions of these equations are

$$E_{\pm,s} = \frac{1}{Y}\left(1 \pm \sqrt{1 - Y}\right) + \mathcal{O}(\varepsilon), \qquad D_s = Y/2E_{\pm,s} + \mathcal{O}(\varepsilon),$$

$$P_s = Y/2 + \mathcal{O}(\varepsilon)$$

$$(4.57)$$

A linear stability analysis of these steady solutions gives the characteristic equation

$$\varepsilon\lambda^3 + 3\varepsilon\lambda^2 + (2D_s + 3\varepsilon + \varepsilon E_{\pm,s}^2)\lambda + \varepsilon + 2D_s + \varepsilon E_{\pm,s}^2 - E_{\pm,s}Y = 0$$

$$(4.58)$$

The three roots of this polynomial are easily found in the form of a power series

$$\lambda_1 = -1 + E_{\pm,s}^2 + \mathcal{O}(\varepsilon^{1/2}) \qquad (4.59)$$

$$\lambda_{2,3} = \pm i \sqrt{Y/(\varepsilon E_{\pm,s})} - 1 - E_{\pm,s}^2/2 + \mathcal{O}(\varepsilon^{1/2}) \qquad (4.60)$$

Close to the limit point, $E_{\pm,s} = 1 \pm \mathcal{O}(\varepsilon)$ and $Y = 1 - \mathcal{O}(\varepsilon)$. Therefore

$$E = E_{+,s}: \quad \lambda_1 > 0, \quad \mathrm{Re}(\lambda_{2,3}) < 0$$
$$E = E_{-,s}: \quad \lambda_1 < 0, \quad \mathrm{Re}(\lambda_{2,3}) < 0$$
$$(4.61)$$

This result states, again, that for the full hysteresis, the intermediate branch is unstable and the lower branch is stable in the absorptive limit. What is interesting is the structure of (4.60), which is associated with damped oscillations. The limit $\varepsilon \to 0$ produces two well separated time scales. The oscillation frequency is $\mathcal{O}(1/\varepsilon^{1/2})$, and the damping is $\mathcal{O}(1)$. The natural way to proceed with this information is to make a multiple time scale analysis by seeking solutions that depend on two times, the time t associated with the damping and the time $T = t/\varepsilon^{1/2}$ associated with the oscillations. This procedure does not work because there is one more element missing in this description. To understand where the additional difficulty lies, we consider equation (4.55) for the field E. The homogeneous part of this equation is $E'' + 2E' + (1 + 2D/\varepsilon)E = 0$. The transformation $E(t) = F(t)\exp(-t)$ leads to an equation that looks like that of the harmonic oscillator $F'' + (2D/\varepsilon)F = 0$. The frequency of this harmonic oscillator is $\omega = \sqrt{2D(t)/\varepsilon}$, which is precisely where we have a problem. Indeed, because D is a function of time, the frequency of the oscillator also varies in time. This analysis suggests that we seek solutions to (4.55)–(4.56) that depend on t and on $T = \omega(t, \varepsilon)/\varepsilon^{1/2}$ in the following way [14, 15]:

$$E(t, \varepsilon) = E(t, T, \varepsilon^{1/2}) = E_0(t, T) + \varepsilon^{1/2}E_1(t, T) + \mathcal{O}(\varepsilon) \qquad (4.62)$$

$$D(t, \varepsilon) = D(t, T, \varepsilon^{1/2}) = D_0(t, T) + \varepsilon^{1/2}D_1(t, T) + \mathcal{O}(\varepsilon) \qquad (4.63)$$

$$\omega(t, \varepsilon) = \omega_0(t) + \varepsilon^{1/2}\omega_1(t) + \mathcal{O}(\varepsilon) \qquad (4.64)$$

Since T is associated with a periodic phenomenon, there remains an undetermination that we remove by requiring that the T-dependent periodic functions be 2π-periodic. For the differentiations, we use the chain rule and the notation

$$dZ/dt = \partial Z/\partial t + (\partial Z/\partial T)(\partial T/\partial t) = Z_t + \varepsilon^{-1/2}\omega'Z_T \qquad (4.65)$$

Inserting the expansions (4.62)–(4.65) into the evolution equations (4.55)–(4.56) leads to

$$\omega'^2 E_{0TT} - Y + 2D_0 E_0 = \mathcal{O}(\varepsilon^{1/2}) \qquad (4.66)$$

$$\varepsilon^{-1/2}\omega'D_{0T} + D_{0t} + \omega'D_{1T} + D_0 - 1 + YE_0/2 = \mathcal{O}(\varepsilon^{1/2}) \qquad (4.67)$$

Equating to zero the coefficients of each power of ε leads to a sequence of simpler problems.

The $\mathcal{O}(\varepsilon^{-1/2})$ problem is $D_{0T} = 0$, which simply states that D_0 depends only on the time t. The $\mathcal{O}(1)$ problem is

$$\omega'^2 E_{0TT} + 2D_0 E_0 = Y \qquad (4.68)$$

$$\omega'D_{1T} = 1 - D_{0t} - D_0 - YE_0/2 \qquad (4.69)$$

To solve (4.68) we set $\omega'^2 = 2D_0$. This specifies the time T and yields a solution to (4.68)

$$T = \int_0^t \sqrt{2D_0(s)/\varepsilon}\,ds \qquad (4.70)$$

$$E_0(t, T) = Y/2D_0(t) + 2\alpha(t)\cos(T) \qquad (4.71)$$

From this result it follows that (4.69) becomes

$$\omega'D_{1T} + Y\alpha(t)\cos(T) = 1 - D_{0t} - D_0 - Y^2/4D_0 \qquad (4.72)$$

The right-hand side of this equation is independent of T. Hence, the integration over T produces a linear divergence in T for all values of the control parameter. This is not physically admissible, and therefore we require that the right-hand side of (4.72) identically vanishes. This gives a closed equation for the variable D_0:

$$D_{0t} = 1 - D_0 - Y^2/4D_0 \qquad (4.73)$$

Finally, to determine fully the first-order solution, we need to find the function $\alpha(t)$. This is not difficult, though the calculation is lengthy. The function $\alpha(t)$ is determined by analyzing the equations at $\mathcal{O}(\varepsilon^{1/2})$ and requiring the boundedness of the solutions. This leads in a straightforward way to

$$\alpha' = -\frac{\alpha}{4D_0}\left(4D_0 + D_0' + Y^2/2D_0\right) \qquad (4.74)$$

Thus we find that for $1/C \to 0$, $\kappa = \mathcal{O}(1)$ and $\gamma = \mathcal{O}(1)$, the problem of solving (4.5)–(4.7) is reduced, in the vicinity of the lower limit point, to a 1-D equation. The active variable is the population difference, whereas the cavity field is the passive variable that simply follows the active variable. The surprise is that (4.73) is the equation that arises in the bad cavity limit when the scaling (4.51) is used. If we expand D_0 around its value at the limit point, we obtain a parabolic equation that can be reduced to the generic form (4.26). The prominent role of the population difference is due to the fact that in the lower branch of the bistable domain, there is little energy in the field. Most of the energy is stored in the material medium and therefore its dynamical properties control the response of the system to external perturbations. This is confirmed by the fact that if we study the vicinity of the limit point on the upper branch in the (E, E_i) plane, we find that the active variable is the field amplitude, and the population difference is the passive variable. Let us just quote the results of the analysis of the upper limit point without details; the procedure is similar to that explained in this section. The scaling of the upper limit point for $1/C = \varepsilon \to 0$ suggests that we seek solutions of the form

$$E = \varepsilon^{-1/2}e, \qquad P = \varepsilon^{1/2}p, \qquad D = \varepsilon d, \qquad E_i = \varepsilon^{-1/2}y \quad (4.75)$$

where the functions e, p, d, and y are $\mathcal{O}(1)$ functions that we expand in powers of $\varepsilon^{1/2}$. To dominant order, a two-time analysis yields the equations

$$e_{0T} = 0, \; e_{0t} = y - e_0 - 2/e_0 \qquad (4.76)$$

$$d_0 = 2\alpha(t)\cos(T) + (e_0 - e_0')/e_0^3 \qquad (4.77)$$

$$\alpha' = -\alpha(1 + 1/e_0^2) \qquad (4.78)$$

$$T = \varepsilon^{-1/2}\int_0^t e_0(s)\,ds \qquad (4.79)$$

Let us add that if we restore the explicit dependence on the three decay rates, the integrand in (4.79) becomes $(e_0^2\gamma_\perp\gamma_\parallel/\gamma_c^2)^{1/2}$, which is precisely the Rabi frequency induced by a field e_0 applied on a two-level medium. Equation (4.76) reduces to equation (4.17), which we obtained in the good cavity limit when the scaling (4.75) was used. Finally, expanding (4.76) in the vicinity of the limit point produces a quadratic equation that can be reduced to the generic form (4.27). The result of this rather heavy analysis is that it unveils some unexpected properties of optical bistability in the large hysteresis domain for $\kappa \cong \gamma \cong 1$ and close to the limit points:

1. The full 3-D dynamics can be reduced to simple 1-D dynamics.

2. The active variable that drives the whole system's evolution is not the same near the two limit points.

3. Critical slowing down affects the time scale T, which is $\mathcal{O}(\varepsilon^{-1/2})$. However, this scaling applies to oscillation frequencies rather than to relaxation time scales.

4. The oscillation frequencies that scale like $\varepsilon^{-1/2}$ are time-dependent. This explains the need to introduce unusual time scales such as (4.70) and (4.79).

References

[1]   L. A. Lugiato, Theory of optical bistability in *Progress in Optics,* Vol. XXI, p. 69 (North-Holland, Amsterdam, 1984).

[2]   H. M. Gibbs, *Optical Bistability: Controlling Light with Light* (Academic Press, Orlando, 1985).

[3]   S. D. Smith, Optical Bistability, Photonic Logic and the Optical Computer. *Phil. Trans. R. Soc. London A* **313** (1984) 187–451.

[4]   P. Mandel, S. D. Smith, and B. S. Wherrett, From Optical Bistability Towards Optical Computing: The EJOB Project (North-Holland, Amsterdam, 1987).

[5]   F. A. Hopf and P. Meystre, *Opt. Commun.* **29** (1979) 235.

[6]   B. Segard, J. Zemmouri, and B. Macke, *Opt. Commun.* **60** (1986) 323.

[7]   A. T. Rosenberg, L. A. Orozco, and H. J. Kimble, Fluctuations and sensitivity in nonequilibrium systems. W. Horsthemke and D. K. Kondepudi, eds., *Springer Proceedings in Physics,* **1** (1984) 62 (Springer, Heidelberg).

[8]   G. Grynberg and S. Cribier, *L. Physique Lett.* **44** (1983) L-449.

[9]   S. Cribier, E. Giacobino, and G. Grynberg, *Opt. Commun.* **47** (1983) 170.

[10] F. Mitschke, C. Boden, W. Lange, and P. Mandel, *Opt. Commun.* **71** (1989) 385.

[11] J. Y. Bigot, A. Daunois, and P. Mandel, *Phys. Lett. A* **123** (1987) 123.

[12] B. Segard, J. Zemmouri, and B. Macke, *Opt. Commun.* **63** (1987) 339.

[13] T. Erneux and P. Mandel, *Phys. Rev. A* **28** (1983) 896.

[14] J. Kervokian and J. D. Cole, *Perturbation Methods in Applied Mathematics, Appl. Math. Sci.* **34** (Springer, Heidelberg, 1981).

[15] C. M. Bender and S. A. Orszag, *Advanced Mathematical Methods for Scientists and Engineers* (McGraw-Hill, New York, 1978).

5

Optical bistability: Variable input

5.1 Swept parameter

In this section, we consider the response of an optically bistable system when the control parameter is swept across a limit point. To analyze this problem, we use the generic equation (4.26) and assume that the control parameter μ is varied linearly in time:

$$x' = x^2 - 2x + \mu, \qquad \mu = \mu_0 + vt, \quad v > 0, \quad x(0) = x_-(\mu_0) < 1 \tag{5.1}$$

This equation turns out to be simple to integrate. Define $x = -u'/u$ and $u = ze^{-t}$. The function z satisfies the Airy equation

$$d^2z/d\xi^2 + \xi z = 0, \qquad \xi = [\mu(t) - 1]v^{-2/3} \tag{5.2}$$

With this result, we can write the solution of (5.1) for an arbitrary sweep rate v as

$$x(t) = 1 + v^{1/3}\frac{Ai'(-\xi) + \alpha Bi'(-\xi)}{Ai(-\xi) + \alpha Bi(-\xi)} \tag{5.3}$$

where α is determined by the initial condition

$$\alpha = -\frac{[x(0) - 1]Ai(\zeta) - v^{1/3}Ai'(\zeta)}{[x(0) - 1]Bi(\zeta) - v^{1/3}Bi'(\zeta)} \tag{5.4}$$

$$\zeta = -\xi(t = 0) = v^{-2/3}(1 - \mu_0) > 0$$

Let us introduce the small sweep rate limit $\zeta \to +\infty$ and simplify the expression for α by using the asymptotic relations (2.17). With these expansions, we find that α is exponentially small

$$\alpha \sim \exp\left[-\frac{4}{3v}(1 - \mu_0)^{3/2}\right] \tag{5.5}$$

Therefore the solution of the original problem (5.1) becomes

$$x(t) = 1 + v^{1/3} Ai'(-\xi)/Ai(-\xi) + \mathcal{O}(v^{2/3}) \qquad (5.6)$$

Let us analyze the nature of this solution in the following different domains.

In the short time limit, $-\xi(t) \gg 1$, the expansions (2.17) lead to

$$x(t) = 1 - v^{1/3}|\xi|^{1/2} + \mathcal{O}(v^{4/3}) = 1 - \sqrt{1 - \mu(t)} + \mathcal{O}(v^{4/3}) \qquad (5.7)$$

This is typically the result we expect: The time-dependent solution is nothing but the steady state solution in which the constant control parameter is replaced by the time-dependent control parameter. Such a solution is often referred to as the *adiabatic* solution. However, this solution has been derived for $\mu(t) < 1$. If $\mu(t)$ approaches unity, $\xi(t)$ tends to zero and the asymptotic expansion (5.7) no longer holds. This is a manifestation of the critical slowing down that we expect in the vicinity of $\mu = 1$. This consideration can be made more quantitative by the following observations.

There is a characteristic time, t_1, at which $x(t_1) = 1$. This time is related to the first zero of the function $Ai'(-\xi)$ given by $\xi_1 \cong 1.019$. This time corresponds to a value of the control parameter

$$\mu(t_1) = 1 + v^{2/3}\xi_1 \qquad (5.8)$$

This implies that there is a delay because the static limit point is at $\mu = 1$. However, the delay is small and vanishes with the 2/3 power of the sweep rate.

Finally, there is a critical time, t_2, related to the first zero of the Airy function $Ai(-\xi)$ given by $\xi_2 \cong 2.338$. The corresponding value of the control parameter is

$$\mu(t_2) = 1 + v^{2/3}\xi_2 \qquad (5.9)$$

At this point, the solution $x(t)$ diverges and the transition toward the upper branch of the complete hysteresis has taken place.

These results are very different from those derived in Chapter 2, where an $\mathcal{O}(1)$ delay was obtained for the laser first threshold, which is a bifurcation point. This drastic reduction of the delay is coherent with the results of Section 3.2.1, where we showed that even an exponentially small modification of the zero solution may significantly reduce the delay.

The results of this section have been successfully applied to the determination of the hysteresis scaling laws in laser diodes [1]. Other experimental tests of the scaling laws in optically bistable systems have been reported in [2]. An extension of this analysis has been proposed in [3] for a back and forth linear sweep. Let $\mu_r > \mu_0$ be the value of the control parameter at which the forward

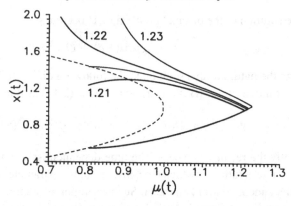

Figure 5.1   Sensitive dependence of the back and forth
sweep on the time at which the sweep changes direction.
Equation (5.1) is integrated with $\mu_0 = 0.8$ and $\nu = 0.1$.
The trajectories are labeled by μ_r. The dashed curve is the
solution of $x^2 - 2x + \mu = 0$. The unlabeled curve is the
dynamical separatrix with $\mu(t_c) \cong 1.216$.

sweep is changed into a backward sweep. Let $x(t_c) = 1$. We use the notation t_c
instead of t_1 as in (5.8) to stress the fact that no small sweep rate assumption has
been introduced. The trajectory associated with $\mu_r = \mu(t_c)$ is the dynamical
separatrix for symmetry reasons. If $\mu_r < \mu(t_c)$, the jump to the upper branch
does not occur. If $\mu_r > \mu(t_c)$, the jump takes place. The trajectory has a very
sensitive dependence on μ_r in the domain $\mu_r \cong \mu(t_c)$. Minute variations of
μ_r produce large variations in the subsequent evolution. This is demonstrated
in Figure 5.1 around the lower limit point. Experimentally, μ_r is always affected by uncontrollable noise, and the result of a large series of experiments
with the same μ_r (apart from the noisy contribution) leads to a family of very
different trajectories. The expression *hesitation phenomenon* has been coined
to describe this effect [3]. The point is that this occurs even for relatively large
sweep rates. However, whatever the sweep rate, $\mu(t_c)$ is given by the exact
relation

$$Ai'(-\xi_c) + \alpha Bi'(-\xi_c) = 0 \qquad (5.10)$$

For small ν, the asymptotic solution (5.8) has the correct scaling as a function
of ν, and the coefficient of ν is also a good approximation. For $\nu = 0.1$ and
$\mu_0 = 0.8$, the asymptotic result (5.8) is $\mu(t_c) \cong 1.2195$, and numerically it is
found to be 1.2160.

Many of the properties described in this section were first found in the study
of OB in the limit of large hysteresis ($C \to \infty$) in the vicinity of the lower limit
point [4].

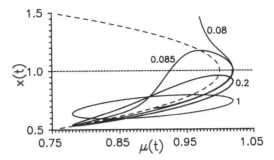

Figure 5.2   Influence of the frequency on the modulation close to a limit point. Equation (5.11) is integrated with $\varepsilon = 0.12$ and $\mu_0 = 0.9$. The curves are labeled by the modulation frequency. The dashed curve is the solution of $x^2 - 2x + \mu = 0$.

5.2 Periodic modulation around a limit point

Another type of time dependence that can be applied in the vicinity of a limit point is a periodic modulation of the control parameter, that is, the input field. The motivation for doing so can be understood qualitatively as follows. Suppose that the modulation amplitude is kept constant and is such that the maximum of the control parameter exceeds the static limit point. If the modulation frequency is sufficiently small, the problem is similar to that of the swept parameter discussed in the previous section and the divergence occurs, corresponding to the up-switching process. However, if the modulation frequency is large enough, the system does not respond to the modulation because of the critical slowing down. In that limit, the periodic control parameter can be replaced by its time average and if that average is less than the static limit point, the system remains in the vicinity of the lower branch. Therefore, there exists a critical frequency separating the two domains. The nice property of this problem is that the stability of the bistable system is controlled by the modulation frequency. This is in contrast with more traditional schemes where the system is controlled by a variation of the amplitude of the control parameter, such as the pulses studied in Sections 4.4.2 and 4.4.3. This property is displayed in Figure 5.2.

Clearly, the modulation amplitude has to be small for this process to happen. Hence we formulate the problem as

$$x' = x^2 - 2x + \mu, \qquad \mu = \mu_0 + \varepsilon \sin(\omega t)$$

$$\mu_0 < 1, \qquad \mu_0 + \epsilon > 1, \qquad x(0) = x_-(\mu_0) \tag{5.11}$$

Using the transformation $x = -u'/u$, we can reduce the problem to a linear second-order differential equation. The bad news, however, is that this time we obtain the Mathieu equation. The solutions of the Mathieu equation are not known in explicit form but only as infinite series. Thus the trick of converting the nonlinear differential equation (5.11) into a linear equation will be of little use and we shall study (5.11) directly by a multiple time scale analysis.

To express the fact that the average value of the control parameter μ_0 is near but below the static limit point, we expand μ_0 in powers of the small parameter ε

$$\mu_0 = 1 - \varepsilon\mu_1 - \varepsilon^2\mu_2 + \mathcal{O}(\varepsilon^3) \tag{5.12}$$

To express the influence of critical slowing down in the vicinity of $(x, \mu) = (1, 1)$, we introduce a slow time scale and seek solutions $x(t, \varepsilon)$ in the form of a series

$$x(t, \varepsilon) = 1 + \varepsilon x_1(t, \tau) + \varepsilon^2 x_2(t, \tau) + \mathcal{O}(\varepsilon^3), \qquad \tau = \varepsilon t \tag{5.13}$$

In the spirit of the multiple time scale analysis, we treat the two times t and τ as independent variables. Using the chain rule $dx/dt = \partial x/\partial t + (\partial x/\partial \tau)(\partial \tau/\partial t) = x_t + \varepsilon x_\tau$, we insert the expansions (5.12) for the control parameter and (5.13) for the solution into the evolution equation (5.11). Equating the coefficient of each power of ε to zero yields a sequence of linear problems.

The first-order problem is

$$x_{1,t} = -\mu_1 + \sin(\omega t) \tag{5.14}$$

whose solution is

$$x_1(t, \tau) = a(\tau) - \mu_1 t - \frac{1}{\omega}\cos(\omega t) \tag{5.15}$$

We have to impose $\mu_1 = 0$ to ensure the boundedness of the solution.

The second-order problem is

$$x_{2,t} = -x_{1,\tau} + x_1^2 - \mu_2 \tag{5.16}$$

and its solution is

$$x_2(t, \tau) = b(\tau) + \left[-\mu_2 - a_\tau + a^2 + \frac{1}{2\omega^2}\right]t + \frac{1}{4\omega^3}\sin(2\omega t) - \frac{2a}{\omega^2}\sin(\omega t) \tag{5.17}$$

Again, we require that the solution $x_2(t, \tau)$ be bounded. This imposes the vanishing of the coefficient of t in the solution (5.17), which gives a closed equation for function $a(\tau)$

$$a_\tau = a^2 - c, \ c = \mu_2 - \frac{1}{2\omega^2} \tag{5.18}$$

There are two possibilities: Either $c > 0$, in which case the steady state $a = -\sqrt{c}$ is the stable solution in the long time limit, or $c < 0$ and the solution $a(\tau)$ diverges in a finite time. If $c > 0$, the solution of (5.18) is

$$a(\tau) = -\sqrt{c} \frac{1 - be^{-2\sqrt{c}\tau}}{1 + be^{-2\sqrt{c}\tau}}, \qquad b = (\sqrt{c} + a_0)/(\sqrt{c} - a_0) \tag{5.19}$$

Hence, for $c > 0$ and in the long time limit $2\sqrt{c}\tau \gg 1$, the solution of equation (5.11) is

$$x(t) = 1 - \varepsilon(\mu_2 - \mu_2^*)^{1/2} - \frac{\varepsilon}{\omega} \cos(\omega t) + \mathcal{O}(\varepsilon^2), \qquad \mu_2^* = 1/2\omega^2 \tag{5.20}$$

$$\mu(t) = 1 + \varepsilon \sin(\omega t) - \varepsilon^2 \mu_2 + \mathcal{O}(\varepsilon^3) \tag{5.21}$$

At the stability boundary, $\mu_2 = \mu_2^*$, the solution becomes

$$x^*(t) = 1 - \frac{\varepsilon}{\omega} \cos(\omega t) + \mathcal{O}(\varepsilon^2) \tag{5.22}$$

$$\mu^* = 1 + \varepsilon \sin(\omega t) - \varepsilon^2/2\omega^2 + \mathcal{O}(\varepsilon^3) \tag{5.23}$$

This result implies that during each cycle, the system spends exactly half its time below the limit point and the other half above the limit point. This is reminiscent of the balance between the accumulated stability and the accumulated instability that explained the delayed bifurcation in Section 2.2.1. Here we see that stability results from a longer time spent below the limit point than above it because there is a systematic shift $-\varepsilon(\mu_2 - \mu_2^*)^{1/2}$.

Note that, to dominant order in ε, the boundary of stability depends on the modulation parameters only via the ratio ε/ω when the solution is expressed in terms of the time τ. Another aspect that should be stressed is that the definition $\mu_2^* \equiv (1 - \mu_0)/\varepsilon^2 = 1/2\omega^2$ establishes a relation between the amplitude of the modulation, its frequency, and the average value of the control parameter at criticality. More generally, for a modulation of amplitude A and frequency ω, and for a limit point at $\mu = \mu_L$, the stability boundary $\mu_2^* \equiv (1 - \mu_0)/\varepsilon^2 = 1/2\omega^2$ becomes

$$A = \omega \sqrt{2(1 - \mu_0/\mu_L)} \tag{5.24}$$

Another point of view that has proved useful in understanding the instability is by reference to the phase lag between the output and the input functions. Using the results (5.15), (5.17), and (5.18), we write the general solution $x(t, \tau)$ as

$$x(t, \tau) = \xi(\tau) + \mathcal{B}(\tau)\cos[\omega t - \phi(\tau)] + \mathcal{C}(\tau)\sin(2\omega t) + \mathcal{O}(\varepsilon^3) \quad (5.25)$$

where $\xi(\tau) = 1 + \varepsilon a(\tau) + \varepsilon^2 b(\tau) + \mathcal{O}(\varepsilon^3)$ is the nonperiodic slowly varying part of the solution and $\mathcal{C}(\tau) = \varepsilon^2/4\omega^3 + \mathcal{O}(\varepsilon^3)$. From this expression it follows that to dominant order in ε we have

$$\mathcal{B}(\tau) = (\varepsilon/\omega)\sqrt{1 + \alpha^2}, \qquad \phi(\tau) = \tan^{-1}(1/\alpha), \qquad \alpha = 2\varepsilon a(\tau)/\omega$$
$$(5.26)$$

In the stable regime, εa is small but positive. Hence the solution $x(t, \tau)$ remains stable as long as the phase lag $\phi(\tau)$ is smaller than $\pi/2$. At criticality, $a = 0$ and therefore the phase lag is exactly $\pi/2$. The relevance of this formulation is that the phase difference between the input and the output variables may be easier to measure than the full output amplitude. This was confirmed by careful experiments in [5]. Other experimental studies on this topic where reported in [3] and [6].

5.3 The onset of dispersive bistability

5.3.1 Steady states and linear stability

We have seen in Section 4.3.2 that in the bad cavity limit and in the dispersive limit, the Maxwell–Bloch equations with an injected field can be reduced to

$$n' = -n + I/[1 + (n - \theta)^2] \quad (5.27)$$

This equation is popular in nonlinear optics because most of the earlier experimental studies in OB were done on systems operated in the strongly dispersive regime. It has also often been used in the scientific literature in the context of the optical transistor. Finally, it represents a very simple description of OB that contains both the monostable and the bistable domains. The choice of n for the variable is made because this equation describes the nonlinear (or intensity-dependent) contribution to the refractive index.

The steady state solution of (5.27) is

$$n^3 - 2n^2\theta + n(\theta^2 + 1) - I = 0 \quad (5.28)$$

From this equation we determine the slope of n versus I

$$dn/dI = \frac{1}{3(n - n_-)(n - n_+)}, \qquad n_\pm = \frac{1}{3}\left(2\theta \pm \sqrt{\theta^2 - 3}\right) \quad (5.29)$$

Thus, if $\theta^2 > 3$, there is bistability and the limit points are located at $n_\pm$. The so-called optical transistor (OT) is obtained in the domain $\theta^2 \leqslant 3$. From the relation

$$d^2n/dI^2 = -6(dn/dI)^3(n - 2\theta/3) \qquad (5.30)$$

we find that if $dn/dI \neq 0$, the maximum slope occurs at

$$n^* = 2\theta/3, \qquad I^* = 2\theta(1 + \theta^2/9)/3 \qquad (5.31)$$

where the slope is $1/(1 - \theta^2/3)$. It becomes infinite for $\theta^2 = 3$ at

$$n_c = 2/\sqrt{3}, \qquad I_c = n_c^3 \qquad (5.32)$$

Finally if we combine the steady state equation (5.28) and the quadratic $3n^2 - 4n\theta + \theta^2 + 1 = 3(n - n_-)(n - n_+) = 0$, we find that the intensity at the limit points is $I_\pm = 2n_\pm^2(\theta - n_\pm)$.

To determine the linear stability of the steady state solutions, we seek solutions of (5.27) in the form $n(t) = n + \varepsilon n_1 + \mathcal{O}(\varepsilon^2)$ and we linearize the equation with respect to ε. The result is $n_1(t) = n_1(0)\exp(\lambda t)$ with

$$\lambda = -1 + 2n^2(\theta - n)/I \qquad (5.33)$$

This characteristic root vanishes at the limit points $(n_\pm, I_\pm)$ and therefore also at the point of infinite slope (n_c, I_c). Consequently, the vicinity of the point of infinite slope is characterized by critical slowing down as well, and we can expect a modification of the dynamical response. In this case, vicinity occurs in phase space and in parameter space. Indeed, the coordinates of the critical point are $(n, I, \theta) = \left(2/\sqrt{3}, (2/\sqrt{3})^3, \sqrt{3}\right)$. Hence the vicinity of the critical point extends in a 2-D phase subspace (n, I) and in a 1-D parameter subspace.

5.3.2 Periodic modulation: The optical transistor

In this section, we analyze the response of an optical transistor (OT) to a periodic modulation of the refractive index. Physically, the OT is biased with one beam of constant intensity I, and the refractive index n is driven by another light source that is amplitude modulated. The biasing beam induces a coherent interaction with the material sample while the amplitude modulated beam induces an incoherent excitation of the medium. For instance, the biasing beam can be produced by a laser with a frequency close to the resonant atomic frequency while the modulated beam is produced by a broadband laser, such as a diode laser emitting far from the atomic frequency. The problem to be solved is therefore

$$n' = -n + I/[1 + (n - \theta)^2] + a\cos(\omega t) \qquad (5.34)$$

Let us begin with the naive assumption that in the limit of a small modulation amplitude, $a = \varepsilon \ll 1$, we can seek a solution of the nonlinear equation (5.34) of the form

$$n(t) = n + \varepsilon g \cos(\omega t + \varphi) + \mathcal{O}(\varepsilon^2) \qquad (5.35)$$

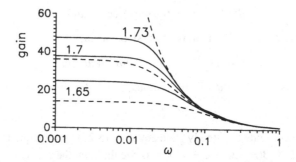

Figure 5.3 The gain of an optical transistor (output over input modulation amplitude) versus the modulation frequency. The solid lines are obtained by solving numerically (5.34). The dashed lines are given by the linear gain (5.37). The curves are labeled by θ. The input intensity is $I = 2\theta(1 + \theta^2/9)/3$ and the modulation amplitude is 0.01. The maximum of the linear gain for $\theta = 1.73$ is $g = 208.9$.

where n is the steady state solution for $\varepsilon = 0$ and g is, by construction, the gain of the transistor. Inserting (5.35) into (5.34) and linearizing with respect to ε leads to the equation

$$(i\omega + 1)ge^{i\varphi} - 1 + \frac{2n^2(n - \theta)}{I}ge^{i\varphi} = 0 \qquad (5.36)$$

Separating real and imaginary parts and eliminating φ among the two equations leads to an expression for the gain

$$1/g(\omega) = \sqrt{\omega^2 + [1 - 2n^2(\theta - n)/I]^2} = \sqrt{\omega^2 + \lambda^2} \qquad (5.37)$$

in terms of the characteristic root (5.33). The difficulty with this result is best appreciated if we consider the maximum of each gain curve, $g(0) = 1/\lambda$, and among that family the value $g^*(0)$ at the point of maximum slope (n^*, I^*). In the monostable case we have

$$g(0) = \frac{I}{I + 2n^2(n - \theta)}, \qquad g^*(0) = \frac{1 + \theta^2/9}{1 - \theta^2/3} \qquad (5.38)$$

Thus, the maximum of the gain $g^*(0)$ is proportional to the slope dn/dI at (n^*, I^*). It diverges if the slope is vertical. Therefore the linear determination of the gain is invalid around (n_c, I_c) and a nonlinear analysis is required. The need for this nonlinear analysis is clearly seen in Figure 5.3, where the linear gain is compared with the gain obtained by integrating numerically equation (5.34). We could approach the nonlinear determination of the gain via a multiple time scale analysis similar to that performed in Section 5.2, as suggested by the occurrence of critical slowing down. However, we use another method that is well adapted for this kind of problem. We assume that equation (5.34) has a solution of the form

$$n(t) = \langle n \rangle + aG\cos(\omega t + \varphi) + \varepsilon R \tag{5.39}$$

Note that this time we do not impose, a priori, a restriction on the modulation amplitude. However, the restriction is that the solution (5.39) captures all the response that oscillates at the external modulation frequency, whereas the harmonics, contained in R, are assumed to be small. In electronics, R is called the *harmonic distortion,* for obvious reasons. The simplest way to proceed is to write (5.34) as $I = [1 + (n - \theta)^2][n' + n - a\cos(\omega t)]$ and expand this equation in powers of ε. To dominant order (i.e., for $\varepsilon = 0$), this yields the pair of equations

$$2G\langle n \rangle (\langle n \rangle - \theta) + [1 + (\langle n \rangle - \theta)^2 + a^2 G^2/2][(1 + i\omega)G - e^{-i\varphi}] = 0 \tag{5.40}$$

$$I = \langle n \rangle [1 + (\langle n \rangle - \theta)^2 + a^2 G^2/2] + a^2 G(\langle n \rangle - \theta)[G - \cos(\varphi)] \tag{5.41}$$

Separating the real and imaginary parts of (5.40) and eliminating the phase φ leads to an equation for the gain

$$1/G(\omega) = \sqrt{\omega^2 + [1 + H(\omega)]^2}, \qquad H(\omega) = \frac{2\langle n \rangle (\langle n \rangle - \theta)}{1 + (\langle n \rangle - \theta)^2 + a^2 G^2(\omega)/2} \tag{5.42}$$

The remaining two equations determine $\langle n \rangle$ and φ. The maximum gain $G(0)$ can be written in terms of the function $F(x) = x[1 + (x - \theta)^2]$ as

$$G(0) = \frac{F(\langle n \rangle) + a^2 G^2(0)/2}{\langle n \rangle [F'(\langle n \rangle) + a^2 G^2(0)/2]}$$

$$\cong \frac{F(\langle n \rangle)}{\langle n \rangle [F'(\langle n \rangle) + a^2 G^2(0)/2]} \tag{5.43}$$

The approximation $F(\langle n \rangle) \gg a^2 G^2(0)/2$ is not simple to justify because $\langle n \rangle$ is not yet determined. Nevertheless, we introduce this simplification and leave it to the reader to verify that the results we obtain are consistent with equations (5.40) and (5.41). Two cases have to be distinguished.

First, if $F'(\langle n \rangle) \gg a^2 G^2(0)/2$, we seek a solution of the form $\langle n \rangle = n + \alpha a^2 G^2(0) + \mathcal{O}[a^4 G^4(0)]$. Inserting this expression into (5.41) and using the real part of (5.40) leads in a straightforward way to the expression for the coefficient α

$$\alpha = \frac{1}{2I} \frac{4n^2(n-\theta)^2 - nI}{(3n-\theta)(n-\theta)+1} \tag{5.44}$$

From this result it follows that at the point of maximum slope we have $\alpha^* = -\theta/3$ and $\langle n \rangle = (2\theta/3)\{1 - a^2G^2(0)/2 + \mathcal{O}[a^4G^4(0)]\}$. Hence the dominant contribution to the maximum gain remains $g^*(0)$ as given in the linear theory by the expression (5.38). However, the domain of validity of this result is now limited by the condition $F'(\langle n \rangle) \gg a^2G^2(0)/2$. Inserting the expansion we have obtained for $\langle n \rangle$ in this inequality leads to the condition

$$a^2 \ll a_c^2 \equiv 2(1 - \theta^2/3)^3/(1 + \theta^2/9) \tag{5.45}$$

The essential implication of this condition is that the linear theory is *not* applicable for finite amplitude if $\theta^2 \to 3$.

Second, if $F'(\langle n \rangle) \ll a^2G^2(0)/2$, the calculation is somewhat simpler since the maximum gain is given by

$$G(0) = \frac{F(\langle n \rangle)}{\langle n \rangle a^2 G^2(0)/2} = \frac{2}{a^2 G^2(0)}\left[1 + (\langle n \rangle - \theta)^2\right] \tag{5.46}$$

and therefore we obtain the expression

$$G(0) = a^{-2/3}2^{1/3}\left[1 + (\langle n \rangle - \theta)^2\right]^{1/3} \tag{5.47}$$

This result is not complete because $\langle n \rangle$ is not yet determined. However, its determination is not necessary for understanding the salient features of the nonlinear gain (5.47) that hold in the domain $a \gg a_c$ complementary to (5.45). This is because it can safely be assumed that $\langle n \rangle$ remains close to the steady state solution n given by $I = F(n)$ as $a \to 0$. For the three values of θ used in Figure 5.3, where $a = 0.01$, the parameter a_c is

θ	1.65	1.70	1.73
a_c	0.1146	0.0451	0.0029

Thus the condition $a \ll a_c$ is satisfied for $\theta = 1.65$ only. The important property to note is that $G(0)$ remains finite as the slope of the OT diverges. More important is that now the gain is a function of the modulation amplitude, quite a natural property: It is the independence of $G(0)$ on a in the linear theory that should have bothered us.

To close this chapter, let us stress that in the notation (5.35) adopted for the linear stability analysis, the result (5.47) means that the maximum gain scales like $g \sim \varepsilon^{-2/3}$. This nonanalytic dependence of the gain on the modulation amplitude explains the failure of the linear response theory.

References

[1] P. Jung, G. Gray, R. Roy, and P. Mandel, *Phys. Rev. Lett.* **65** (1990) 1873.

[2] J. Grohs, H. Issler, and C. Klingshirn, *Opt. Commun.* **86** (1991) 183.

[3] J. Zemmouri, B. Ségard, W. Sergent, and B. Macke, *Phys. Rev. Lett.* **70** (1993) 1135.

[4] T. Erneux and P. Mandel, *Phys. Rev. A* **28** (1983) 896.

[5] C. Boden, F. Mitschke, and P. Mandel, *Opt. Commun.* **76** (1990) 178.

[6] M. James and F. Moss, *J. Opt. Soc. Am. B* **5** (1988) 1121.

6

Multimode optical bistability

The title of this chapter may be misleading and needs a warning because there are two ways in which multimode operation can be studied in the framework of the Maxwell–Bloch equations. The first approach is to project the field E onto an orthonormal basis of modes. However, many problems are related to this procedure. The main difficulty is that one does not know how to define and construct a complete and orthogonal basis of eigenmodes for a lossy cavity. Thus the basis must be associated with an ideal (or lossless) cavity. It is not known to what extent some properties, if any, are lost in this way because in many respects the lossless cavity is the singular limit of the lossy cavity. However, the use of a modal decomposition has proved itself useful for the study of few mode problems. The second approach is to work directly with the partial differential equations without using modal expansions.

In this chapter and in Chapter 12, we analyze multimode problems of non-linear optics for which a global approach is more adequate. Modal expansions will be used in Chapters 7 and 10.

6.1 The delay-differential equation

6.1.1 Reduction of the Maxwell–Bloch equations

We start with the Maxwell–Bloch equations (1.25)–(1.27) and the boundary condition (1.30). The field equation (1.25) can be written as

$$c^2 \frac{\partial}{\partial z}\left(2k_c - i\frac{\partial}{\partial z}\right)E_0 + \frac{\partial}{\partial t}\left(2\omega_c + i\frac{\partial}{\partial t}\right)E_0 = -\frac{N\mu}{\varepsilon_0}\left(\omega_c + I\frac{\partial}{\partial t}\right)^2 P_0 \quad (6.1)$$

We first perform a simplification based on the assumption that E_0 and P_0 are slowly varying on the optical space–time scale

$$\omega_c|P_0| \gg |\partial P_0/\partial t|, \qquad \omega_c|E_0| \gg |\partial E_0/\partial t|, \qquad k_c|E_0| \gg |\partial E_0/\partial z| \quad (6.2)$$

66

We have shown in Chapter 1 how this "fast physics" can be justified. With these approximations, we get the following equations.

$$c\, \partial E_0/\partial z + \partial E_0/\partial t = -(N\mu\omega_c/2\varepsilon_0)P_0 \tag{6.3}$$

$$\partial P_0/\partial t = -i(\omega_a - \omega_c)P_0 - (\mu/\hbar)E_0 D - \gamma_\perp P_0 \tag{6.4}$$

$$\partial D/\partial t = (\mu/2\hbar)(P_0^* E_0 + P_0 E_0^*) - \gamma_\parallel(D - D_a) \tag{6.5}$$

$$E_0[t, 0] = \sqrt{T_f} E_i + R E_0[t - (L - \ell)/c, \ell] \tag{6.6}$$

The boundary condition arises from the boundary for the real fields $E[t, 0] = \sqrt{T_f} E_i + RE[t - (L - \ell)/c, \ell]$ and the assumption that the cavity field and the injected field oscillate at the same frequency. We recall that L is the cavity length and ℓ is the length of the nonlinear medium. An additional quantity that will be useful is the field transmitted through the out-coupling mirror placed just after the nonlinear medium

$$E_{\text{out}}(t) = \sqrt{T_f} E(t, \ell) \tag{6.7}$$

Because we are dealing with a passive system, the population inversion D_a is negative. We introduce $P = -P_0/D_a$ and $D = -D/D_a$. The new atomic variables satisfy the equations

$$\partial P/\partial t = -i(\omega_a - \omega_c)P - (\mu/\hbar)E_0 D - \gamma_\perp P \tag{6.8}$$

$$\partial D/\partial t = (\mu/2\hbar)(P^* E_0 + P E_0^*) - \gamma_\parallel(D + 1) \tag{6.9}$$

In Chapter 1, we made the change of space–time variables (1.31) to remove the delay appearing in the boundary condition. In this chapter, the attitude is the opposite: We exploit the unavoidable delay due to the finite propagation velocity of light. Hence we introduce the new space–time variables

$$\tau = t - z/c, \qquad \zeta = z \tag{6.10}$$

and $Z(t, z) \to Z(\tau, \zeta)$ where Z is any of the three functions E_0, P, and D. Using the chain rule for differentiation leads to the relations $\partial f/\partial t = \partial f/\partial\tau$ and $\partial f/\partial z = \partial f/\partial\zeta - (1/c)\partial f/\partial\tau$. Therefore, the evolution equations (6.3, 6.8, and 6.9) become

$$\partial E_0/\partial\zeta = (N\mu k_c D_a/2\varepsilon_0)P \tag{6.11}$$

$$\partial P/\partial\tau = -[i(\omega_a - \omega_c) + \gamma_\perp]P - (\mu/\hbar)E_0 D \tag{6.12}$$

$$\partial D/\partial\tau = (\mu/2\hbar)(P^* E_0 + P E_0^*) - \gamma_\parallel(D + 1) \tag{6.13}$$

At this point, we introduce the usual assumption that leads to the adiabatic elimination of the atomic polarization: We assume that P relaxes much faster than both the field and the population difference. Thus we approximate (6.12) by

$$P(\tau, \zeta) \cong -(\mu/\hbar)E_0(\tau, \zeta)D(\tau, \zeta)/[i(\omega_a - \omega_c) + \gamma_\perp] \qquad (6.14)$$

Inserting this relation into (6.11) leads to

$$E_0(\tau, \zeta) = E_0(\tau, 0) \exp\left(\beta \overline{D} \frac{\gamma_\perp - i\Omega}{\gamma_\perp^2 + \Omega^2}\right) \qquad (6.15)$$

with

$$\overline{D} = \int_0^\zeta D(\tau, \zeta')\, d\zeta' \equiv \overline{D}(\tau, \zeta),$$

$$\beta = -N\mu^2 k_c D_a/2\hbar\varepsilon_0, \qquad \Omega = \omega_a - \omega_c \qquad (6.16)$$

With this result, we manipulate (6.13) as follows

$$\partial D/\partial \tau = -\gamma_\parallel(D + 1) - 2(\mu/2\hbar)^2 \left(\frac{\gamma_\perp - i\Omega}{\gamma_\perp^2 + \Omega^2} + \frac{\gamma_\perp + i\Omega}{\gamma_\perp^2 + \Omega^2}\right) D|E_0|^2$$

$$= -\gamma_\parallel(D + 1) - (\mu/\hbar)^2 \frac{\gamma_\perp}{\gamma_\perp^2 + \Omega^2} D|E_0|^2$$

$$= -\gamma_\parallel(D + 1) - (\mu/\hbar)^2 \frac{\gamma_\perp}{\gamma_\perp^2 + \Omega^2} D|E_0(\tau, 0)|^2 \exp\left(\frac{2\beta\gamma_\perp \overline{D}}{\gamma_\perp^2 + \Omega^2}\right)$$

$$= -\gamma_\parallel(D + 1) - \mu^2/(2\beta\hbar^2)|E_0(\tau, 0)|^2 \frac{\partial}{\partial \zeta} \exp\left(\frac{2\beta\gamma_\perp \overline{D}}{\gamma_\perp^2 + \Omega^2}\right) \qquad (6.17)$$

Finally, integrating this equation with respect to ζ gives

$$\partial \overline{D}/\partial \tau = -\gamma_\parallel(\overline{D} + \zeta) - \mu^2/(2\beta\hbar^2)|E_0(\tau, 0)|^2 \left[\exp\left(\frac{2\beta\gamma_\perp \overline{D}}{\gamma_\perp^2 + \Omega^2}\right) - 1\right] \qquad (6.18)$$

We define

$$\Phi(\tau) = 1 + (1/\ell)\overline{D}(\tau - \ell/c, \ell), \qquad \alpha = \frac{\beta\gamma_\perp}{\gamma_\perp^2 + \Omega^2} \qquad (6.19)$$

and rescale all electric fields (input field, cavity field, and transmitted field) according to

$$\mathcal{E} = (\mu/2\hbar)E\left(\gamma_\parallel\gamma_\perp(1 + \Omega^2/\gamma_\perp^2)\right)^{-1/2} \qquad (6.20)$$

With these definitions, we obtain for the function $\Phi(\tau)$ the nonlinear equation

$$d\Phi(\tau)/d\tau = -\gamma_\parallel\Phi(\tau) - \frac{2\gamma_\parallel}{\alpha\ell}|\mathcal{E}(\tau - \ell/c, 0)|^2 \left(e^{2\alpha\ell[\Phi(\tau)-1]} - 1\right) \qquad (6.21)$$

and for the field amplitude

$$\mathcal{E}(\tau, \ell) = \mathcal{E}(\tau - \ell/c, 0) \exp\{\alpha\ell(1 - i\Omega/\gamma_\perp)[\Phi(\tau) - 1]\} \qquad (6.22)$$

Using this result, we write the boundary conditions (6.6) and (6.7) as

$$\mathcal{E}_{\text{out}}(\tau) = \sqrt{T_f}\mathcal{E}(\tau - \ell/c, 0) \exp\{\alpha\ell(1 - i\Omega/\gamma_\perp)[\Phi(\tau) - 1]\} \quad (6.23)$$

and

$$\mathcal{E}(\tau, \ell) = \left(\sqrt{T_f}\mathcal{E}_i + R\mathcal{E}(\tau - \tau_r, \ell)\right)\exp\{\alpha\ell(1 - i\Omega/\gamma_\perp)[\Phi(\tau) - 1]\} \quad (6.24)$$

where we have introduced the cavity round-trip time $\tau_r = L/c$.

The set of equations (6.21), (6.23), and (6.24) form a convenient starting point for studying those aspects of optical bistability that depend on the finite velocity of light and require a description incorporating explicitly the round-trip time of the light in the cavity. We proceed in two steps. First we derive from (6.21), (6.23), and (6.24) a closed delay-differential equation for the variable $\Phi(\tau)$. Afterward, we reduce this equation further and obtain a discrete map.

To simplify (6.21), we consider the dispersive limit in which we neglect the effect of the nonlinear absorption. Starting with (6.21), we have

$$d\Phi(\tau)/d\tau \cong -\gamma_\parallel\Phi(\tau) + \frac{2\gamma_\parallel}{\alpha\ell}|\mathcal{E}(\tau - \ell/c, 0)|^2\left(1 - e^{-2\alpha\ell}\right) \quad (6.25)$$

Inverting (6.23) and neglecting the nonlinear absorption, we introduce the transmitted field in (6.25)

$$d\Phi(\tau)/d\tau = -\gamma_\parallel\Phi(\tau) + \frac{2\gamma_\parallel}{\alpha\ell T_f}|\mathcal{E}_T(\tau)|^2\left(e^{2\alpha\ell} - 1\right) = -\gamma_\parallel\Phi(\tau) + \gamma_\parallel|\overline{\mathcal{E}}(\tau)|^2$$
$$(6.26)$$

with

$$\overline{\mathcal{E}}(\tau) \equiv \mathcal{E}_{\text{out}}(\tau)\sqrt{2\frac{e^{2\alpha\ell} - 1}{\alpha\ell T_f}} \qquad (6.27)$$

Combining (6.23) and (6.24) and using (6.27) to rescale the injected field yields the relation

$$\overline{\mathcal{E}}(\tau) = \sqrt{T_f}\left(\sqrt{T_f}\,\overline{\mathcal{E}}_i + \frac{R}{\sqrt{T_f}}\overline{\mathcal{E}}(\tau - \tau_r)\right)\exp\{\alpha\ell(1 - i\Omega/\gamma_\perp)[\Phi(\tau) - 1]\}$$

$$= \left(\overline{\mathcal{E}}_iT_fe^{\alpha\ell[\Phi(\tau)-1]} + \overline{\mathcal{E}}(\tau - \tau_r)Re^{\alpha\ell[\Phi(\tau)-1]}\right)\exp\{-i\alpha\ell(\Omega/\gamma_\perp)[\Phi(\tau) - 1]\}$$

$$\cong \left(\overline{\mathcal{E}}_iT_fe^{-\alpha\ell} + \overline{\mathcal{E}}(\tau - \tau_r)Re^{-\alpha\ell}\right)\exp\{-i\alpha\ell(\Omega/\gamma_\perp)[\Phi(\tau) - 1]\} \quad (6.28)$$

We introduce a final scaling

$$\varphi(\tau) = (\alpha\ell\Omega/\gamma_\perp)\Phi(\tau), \qquad \overline{E}_{\text{out}}(\tau) = \sqrt{\alpha\ell\Omega/\gamma_\perp}\,\overline{\mathcal{E}}(\tau) \quad (6.29)$$

In terms of these new variables, the dispersive limit of equations (6.21), (6.23), and (6.24) reads

$$\varphi'(\tau) = \gamma_{\parallel}\left(-\varphi(\tau) + |\overline{E}_{\text{out}}(\tau)|^2\right) \qquad (6.30)$$

$$\overline{E}_{\text{out}}(\tau) = \left(A + B\overline{E}_{\text{out}}(\tau - \tau_r)\right)\exp\{i[\varphi(\tau) - \varphi(0)]\} \qquad (6.31)$$

with the coefficients

$$A = E_i(\mu/\hbar)\sqrt{\frac{\Omega T_f}{2\gamma_{\perp}}\frac{1 - e^{-2\alpha\ell}}{\gamma_{\perp}\gamma_{\parallel}(1 + \Omega^2/\gamma_{\perp}^2)}}, \qquad B = Re^{-\alpha\ell} \qquad (6.32)$$

Let us consider the limit $B \ll 1$. From (6.31), we have

$$|\overline{E}_{\text{out}}(\tau)|^2 = |A + B\overline{E}_{\text{out}}(\tau - \tau_r)|^2 \cong A^2 + 2A^2B\cos[\varphi(\tau - \tau_r) - \varphi_0] + \mathcal{O}(B^2) \qquad (6.33)$$

that leads to a closed equation for the variable $\varphi(\tau)$

$$(1/\gamma_{\parallel})\, d\varphi(\tau)/d\tau = -\varphi(\tau) + A^2\left(1 + 2B\cos[\varphi(\tau - \tau_r) - \varphi_0]\right) \qquad (6.34)$$

This equation is known as the *Ikeda delay-differential equation* [1]. This equation and its 2-D version (6.30)–(6.31) have been the subject of many studies. Unfortunately, the analytical results are rare in the field of delay-differential equations. The main results pertaining to (6.34) and/or (6.30)–(6.31) have been obtained by means of numerical simulations performed in several limits. The first studies were made by Ikeda and his collaborators ([2]–[3]). A recent extension of these studies is summarized in [4]. The first experiments used an electrooptic set-up where the output of the optical system was stored electronically and fed back to the system after a delay ([5]–[7]). The first all-optical realization of the equation (6.34) used an optical fiber as delay line [8]. Finally, a thorough investigation of the 2-D problem (6.30)–(6.31) was reported in [9]. After these initial works, many other systems, such as opto-acoustic devices, were shown to display features similar to the optical system described here. Analytic results and further references can be found in [10] and [11].

Considering the number of parameters that enter in the description of optical bistability, it is easy to imagine that many limits are possible. Among them, however, there is the limit $\gamma_{\parallel}\tau_r \gg 1$ in which a dramatic reduction occurs. This will be considered in detail in Section 6.2.

6.1.2 Linear stability analysis

One of the few results that can be derived analytically is the linear stability of the steady state solutions of (6.34). To reduce the number of parameters, we fix $\varphi_0 = -\pi/2$ and write (6.34) as

$$(1/g)\, d\varphi(T)/dT = -\varphi(T) + a - b\sin[\varphi(T - 1)]$$

$$T = \tau/\tau_r, \qquad g = \gamma_\parallel \tau_r \qquad (6.35)$$

The steady state solutions of this equation satisfy the equation $\varphi = a - b\sin(\varphi)$. A linear stability of this steady solution yields the characteristic equation

$$1 + be^{-\lambda}\cos(\varphi) + \lambda/g = 0 \qquad (6.36)$$

Let us write $\lambda = r + i\omega$. Instabilities occur if $r = 0$. Then from (6.36) we obtain the pair of equations $\cos(\varphi)\cos(\omega) = -1/b$ and $\cos(\varphi)\sin(\omega) = \omega/gb$, from which we deduce the relations

$$1 + (\omega/g)^2 = [b\cos(\varphi)]^2 \qquad (6.37)$$

$$\tan(\omega) = -\omega/g \qquad (6.38)$$

These equations have one real root $\omega = 0$ corresponding to limit points in the steady state solution. Besides this root, there is an infinite number of finite roots $\omega \neq 0$ corresponding to Hopf bifurcations. Out of these bifurcation points, periodic solutions emerge. The first complex root has a frequency ω that varies between π and $\pi/2$ as g varies between ∞ and 0. Thus the first periodic solution emerging from a bifurcation of the steady solution has a period varying between 2 (for $g = \infty$) and 4 (for $g = 0$). With the scaling $T = \tau/\tau_r$, the periods are in units of the round-trip time. In the parameter plane (a, b) the locus of the first Hopf bifurcation is given by

$$a_1 = \pm\cos^{-1}(\Gamma_1/b) + 2\pi k \pm b\sin[\cos^{-1}(\Gamma_1/b)] \qquad (6.39)$$

with $\Gamma_1 = -\sqrt{1 + (\omega/g)^2}$. To close this section, let us give a warning. Equation (6.38) has an infinite number of solutions $\omega = \omega_j$, to which there corresponds an infinite number of bifurcations and therefore an infinite number of boundaries a_j given by (6.39) with Γ_1 replaced by $\Gamma_j = -\sqrt{1 + (\omega_j/g)^2}$. As long as the boundaries remain at a finite distance from each other, this striation of the parameter space is without consequence. Indeed, once the first boundary has been crossed, the solution is periodic and what matters is the stability boundary for the periodic solution. However, in the limit $g \to \infty$ all the boundaries a_j coalesce and this degeneracy signals a singularity. This singularity has been studied by few authors and requires advanced tools of mathematical analysis that will not be considered here.

6.2 The discrete map

6.2.1 Constant control parameter

It is instructive to rewrite the delay-differential equation (6.34) in terms of the reduced time $T = \tau/\tau_r$

$$(1/\gamma_\parallel \tau_r)\, d\varphi(T)/dT \;=\; -\varphi(T) + A^2\big(1 + B\cos[\varphi(T-1) - \varphi_0]\big) \quad (6.40)$$

In the singular limit $\tau_r\gamma_\parallel \to \infty$, (6.40) becomes a purely discrete equation. For convenience, we choose $\varphi_0 = -\pi/2$. In the discrete equation, T is defined only for nonnegative integers and therefore we replace it by n, which is more conventional. This leads to the so-called Ikeda map [12]

$$\varphi(n) = a - b\sin[\varphi(n-1)] \tag{6.41}$$

In view of the comments made in the last section, let us state that it is perfectly legitimate to consider the limit of the map. The difficulty arises when trying to connect the properties of the map with those of the delay-differential equation.

Another limit of (6.40) is obtained by setting $\varphi_0 = 0$ and taking again the singular limit $\tau_r\gamma_\parallel \to \infty$. Expanding the cosine function in powers of its argument leads to the Feigenbaum quadratic map

$$\psi(n+1) = 1 - \mu[\psi(n)]^2 \tag{6.42}$$

with $\varphi(n) = A^2(1+B)\psi(n)$ and $\mu = A^4 B(1+B)/2$. Let us analyze the Ikeda map. It has a fixed point solution $\varphi = a - b\sin(\varphi)$. To study its stability against small perturbations, we seek solutions of the map (6.41) in the form $\varphi(n) = \varphi + \varepsilon\phi e^{\lambda n}$. Hence the asymptotic stability condition will be $|e^\lambda| < 1$ to guarantee that $\varphi(n) \to \varphi$ as $n \to \infty$. Inserting this expression for $\varphi(n)$ into (6.41) and linearizing with respect to ε leads to the characteristic equation

$$b\cos(\varphi) = -\exp(\lambda) \tag{6.43}$$

Let $f(x) \equiv a - b\sin(x)$ and $f'(x) = -b\cos(x)$. We can classify the critical solutions (defined by $\mathrm{Re}(\lambda) = 0$) of the characteristic equation (6.43) in two categories:

1. Tangent boundaries

$$f(\varphi) = \varphi, \qquad f'(\varphi) = +1 \tag{6.44}$$

   At such a boundary, $\lambda = 0$, which signals a limit point in the multistable curve $\varphi = \varphi(a, b)$, that is, the birth of a pair of stable–unstable solutions.

2. Harmonic boundaries

$$f(\varphi) = \varphi, \qquad f'(\varphi) = -1 \tag{6.45}$$

   At these boundaries, $\lambda = \pm i\pi$ (modulo 2π) corresponding to a bifurcation point from which a period 2 solution emerges.

Another remarkable solution of the map is the superstable (period 1) orbit occurring for

$$f(\varphi) = \varphi, \qquad f'(\varphi) = 0 \tag{6.46}$$

In this case $\lambda = -\infty$ and the fixed point solution is located at an extremum of the map.

It is now a simple exercise to obtain the explicit equation for the harmonic and tangent boundaries in the parameter plane (a, b). Let us study in detail the harmonic boundary. From $\cos(\varphi) = 1/b$, it follows that $b > 1$ and either $3\pi/2 \leq \varphi \leq 2\pi$ or $0 \leq \varphi \leq \pi/2$ (modulo 2π).

- In the domain $3\pi/2 \leq \varphi \leq 2\pi$, we have $\cos(\varphi) = \cos(2\pi - \varphi) = 1/b$ and $\sin(\varphi) = -\sqrt{1 - 1/b^2}$. Hence $2\pi - \varphi + 2\pi n = \cos^{-1}(1/b)$ and therefore

$$a = 2\pi(n + 1) - \sqrt{b^2 - 1} - \cos^{-1}(1/b)$$

$$\text{if } \pi(2n + 3/2) \leq \varphi \leq 2\pi(n + 1) \qquad (6.47)$$

- In the domain $0 \leq \varphi \leq \pi/2$, we have $\cos(\varphi) = 1/b$ and $\sin(\varphi) = \sqrt{1 - 1/b^2}$. Hence $\varphi + 2\pi n = \cos^{-1}(1/b)$ and therefore

$$a = 2\pi n + \sqrt{b^2 - 1} + \cos^{-1}(1/b)$$

$$\text{if } 2\pi n \leq \varphi \leq \pi(2n + 1/2) \qquad (6.48)$$

In a similar way, the tangent boundaries are given by

$$a = 2\pi(n + 1) + \sqrt{b^2 - 1} - \cos^{-1}(1/b)$$

$$\text{if } \pi(2n + 1/2) \leq \varphi \leq \pi(2n + 1) \qquad (6.49)$$

$$a = 2\pi(n + 1) - \sqrt{b^2 - 1} - \cos^{-1}(1/b)$$

$$\text{if } \pi(2n + 1) \leq \varphi \leq \pi(2n + 3/2) \qquad (6.50)$$

This completes the analysis of the period 1 solutions of (6.41). The period 2 solutions that emerge from one of the harmonic boundaries (6.47) or (6.48) can be analyzed in the same way because the period 2 solutions of the map (6.41) are the period 1 solutions of the first iterate of the map

$$\varphi(n) = a - b\sin\{a - b\sin[\varphi(n - 2)]\} = f\{f[\varphi(n)]\} \equiv f^{(2)}(\varphi) \quad (6.51)$$

All the analysis we have presented for the fixed point of the Ikeda map (6.41) can be repeated for the fixed points of the iterated map. The characteristic equation is

$$e^{2\lambda} = b^2\cos(\varphi)\cos[a - b\sin(\varphi)] = \frac{d}{d\varphi}\{a - b\sin[a - b\sin(\varphi)]\} \quad (6.52)$$

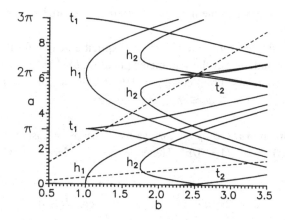

Figure 6.1 Tangent (t_N) and harmonic (h_N) boundaries for the Ikeda map (6.41) in the parameter plane
(a, b). The upper dashed line is $b = 2a/5$. The lower
dashed line is $b = 5a/2$. (After [17].)

The bifurcation boundaries will be the simultaneous solutions of

$$f^{(2)}(\varphi) = \varphi, \qquad f^{(2)\prime}(\varphi) = \pm 1 \tag{6.53}$$

where $+1$ refers to the tangent boundaries ($\lambda = 0$) and -1 refers to the harmonic boundaries ($\lambda = \pm i\pi/2$).

This procedure can be repeated for each iterate of the map, thereby producing
an infinite set of boundaries in the (a, b) plane. As an example, the boundaries
h_1, h_2, t_1, and t_2 have been plotted in Figure 6.1. Note that all the stripes $2n\pi \leq$
$a \leq 2(n + 1)\pi$ where n is a nonnegative integer are identical.

Figure 6.1 contains the domain $0 \leq a \leq 3\pi$ for a better presentation of the
boundary structure. To the left of the h_1 and t_1 boundaries, the fixed point of
$\varphi = a - b\sin(\varphi)$ is stable. Crossing a t_1 boundary implies a limit point and
a local hysteresis. Crossing an h_1 boundary means crossing a period doubling
bifurcation point. As an illustration, we have plotted in Figure 6.2 the solution
of the Ikeda map (6.41) along the line $b = 2a/5$. This line is shown in Figure
6.1. Following this line in the plane (a, b), we see that the solution first crosses
a t_1 boundary twice and then two harmonic boundaries, h_1 and h_2, respectively.
This is indeed the sequence of bifurcations that is seen in Figure 6.2. Another
example is shown in Figure 6.3, corresponding to $b = 5a/2$. In this case only
a pair of h_1 and h_2 are crossed. In either Figure 6.2 or 6.3, the solution enters a
complex sequence of chaotic domains and periodic windows as the parameter
a is further increased.

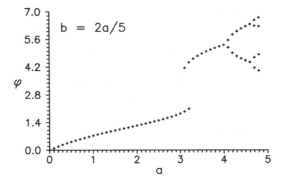

Figure 6.2 Solution of the Ikeda map (6.41) for $b = 2a/5$. The solution crosses twice a tangent boundary t_1 and then the two harmonic boundaries h_1 and h_2.

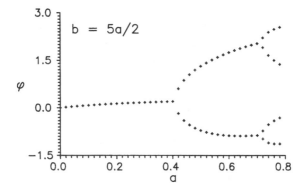

Figure 6.3 Solution of the Ikeda map (6.41) for $b = 5a/2$. The solution crosses the two harmonic boundaries h_1 and h_2.

6.2.2 Swept parameter

In the previous chapters, we have devoted considerable attention to the question of swept parameter. Can we say something in the case of a map with swept parameter? To analyze this question, we consider the limit of small $\varphi(n)$ in the Ikeda map, which then reduces to the Feigenbaum map (6.42). The transformation $\psi(n) = a + 2bx(n)$ applied to (6.42) leads to the more convenient form $x(n+1) = \lambda x(n)[1 - x(n)]$ with $\mu a^2 + a - 1 = 0$, $\lambda = -2\mu a$ and $b = \lambda/\mu$. Since n is the discretized time, the problem with swept parameter can be formulated as

$$x(n + 1) = \lambda(n)x(n)[1 - x(n)], \qquad \lambda(n + 1) = \lambda(n) + v \qquad (6.54)$$

If $v = 0$, $\lambda(n) \equiv \lambda$ and the fixed points of this map are the trivial solution $x = 0$ and the finite solution $x = 1 - 1/\lambda$. The trivial solution is stable for $0 \leq \lambda \leq 1$, whereas the nontrivial fixed point is stable in the domain $1 \leq \lambda \leq 3$. The main point that we must stress here is that $x = 0$ is *always* an exact fixed point of the map, whether λ is time-dependent or not. Let us therefore consider the solution of the nonautonomous map (6.54) with $x(0)$ close to 0 and $0 \leq \lambda(0) \leq 1$. We linearize (6.54) around $x = 0$ and obtain the linear equation $x(n + 1) = \lambda(n)x(n)$ whose solution is

$$x(n) = x(0) \prod_{j=0}^{n-1} \lambda(j) \tag{6.55}$$

The solution remains bounded as long as the coefficient of $x(0)$ in (6.55) is not greater than unity. Thus we can define a critical time, n^*, by the condition

$$\prod_{j=0}^{n^*-1} \lambda(j) = 1 \tag{6.56}$$

To make the connection between the swept parameter problem in the continuous domain and this result, we introduce an intermediate time $\bar{n}$ defined by the condition that $\lambda(\bar{n})$ is closest to unity. Then taking the logarithm of (6.56) leads to the balance equation

$$-\sum_{j=0}^{\bar{n}} \ln[\lambda(j)] = \sum_{j=\bar{n}+1}^{n^*-1} \ln[\lambda(j)] \tag{6.57}$$

In this form, the instability condition appears again as a balance between accumulated stability and accumulated instability. The implicit equation (6.57) is the analog, for a discrete problem, of the condition (2.12) derived for differential equations. The solution of the nonautonomous map is displayed in Figure 6.4 in a domain that surrounds the first bifurcation. It is seen on this figure that we recover practically the result of Chapter 2, namely $\lambda(n^*) - \lambda(\bar{n}) = \lambda(\bar{n}) - \lambda(0)$.

In the limit of a small sweep rate, we can approximate the sums by integrals in (6.57) and obtain the equation

$$\lambda^*[1 - \ln(\lambda^*)] = \lambda_0[1 - \ln(\lambda_0)] = \alpha, \qquad \lambda^* \equiv \lambda(n^*) > 1, \qquad \lambda_0 \equiv \lambda(0) < 1 \tag{6.58}$$

Since $\alpha < 1$, the explicit solution of (6.58) is

$$\lambda^* = \exp\left(1 - \sum_{j=1}^{\infty} \frac{j^{j-1}}{j!}(\alpha/e)^j\right) \tag{6.59}$$

as demonstrated in [13]. Thus, λ^* decreases monotonically from $\lambda^* = e$ (maximum delay) if $\lambda_0 = 0$ to $\lambda^* = 1$ (no delay) if $\lambda_0 = 1$.

An analysis of the delay at later bifurcations is a much more complex problem that requires the methods of either the renormalization group or the nonstandard

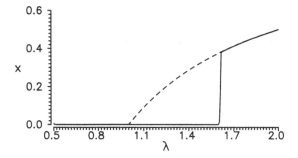

Figure 6.4 Solution of the nonautonomous map (6.54) for $v = 0.001$. The initial condition is $x(0) = 0.01$ and $\lambda(0) = 0.5$. The dashed line is the function $x = \lambda x(1 - x)$ given for reference.

group theory. This is no longer within the scope of this book and the reader is referred to the original publications, [14] and [15], for more details. Analog simulations were reported in [16].

References

[1] K. Ikeda, *Opt. Commun.* **30** (1979) 257.

[2] K. Ikeda and O. Akimoto, *Phys. Rev. Lett.* **48** (1982) 617.

[3] K. Ikeda, K. Kondo, and O. Akimoto, *Phys. Rev. Lett.* **49** (1982) 1467.

[4] K. Ikeda, K. Otsuka, and K. Matsumoto, *Progress of Theor. Phys.* **99** (1989) 295.

[5] H. M. Gibbs, F. A. Hopf, D. L. Kaplan, and R. L. Shoemaker, *Phys. Rev. Lett.* **46** (1981) 474.

[6] F. A. Hopf, D. L. Kaplan, H. M. Gibbs, and R. L. Shoemaker, *Phys. Rev. A* **25** (1982) 2172.

[7] M. W. Derstine, H. M. Gibbs, F. A. Hopf, and D. L. Kaplan, *Phys. Rev. A* **26** (1982) 3720; *ibid.* **27** (1983) 3200.

[8] H. Nakatsuka, S. Asaka, H. Itoh, K. Ikeda, and M. Matsuoka, *Phys. Rev. Lett.* **50** (1983) 109.

[9] M. Le Berre, E. Ressayre, A. Tallet, and H. M. Gibbs, *Phys. Rev. Lett.* **56** (1986) 274; M. Le Berre, E. Ressayre, and A. Tallet, *Phys. Rev. A* **41** (1990) 6635.

[10] S. N. Chow and J. Mallet-Paret in Coupled nonlinear oscillators, J. Chandra and A. C. Scott, eds. (North-Holland, Amsterdam, 1983).

[11] P. Nardone, P. Mandel, and R. Kapral, *Phys. Rev. A* **33** (1986) 2465.

[12] K. Ikeda, H. Daido, and O. Akimoto, *Phys. Rev. Lett.* **45** (1980) 709.

[13] L. Comtet, *Advanced Combinatorics: The Art of Finite and Infinite Expansions*, Chap. 5, Exercise 22 (D. Reidel, Dordrecht, 1974).

[14] A. Fruchard, *C. R. Acad. Sci. Paris* **307** (1988) 41.

[15] C. Baesens, *Physica D* **53** (1991) 319.

[16] B. Morris and F. Moss, *Phys. Lett. A* **118** (1986) 117.

[17] P. Mandel and R. Kapral, *Opt. Commun.* **47** (1983) 151.

7

Free-running multimode lasers

In Chapters 1 to 5, we have dealt with single-mode ring cavities, either for lasers or for optical bistability. In this chapter, we come back to laser theory to consider the properties of multimode cavities. This subject is immense and our goal can only be modest.

The single-mode unidirectional ring laser is the model of choice for theoreticians who want to study fundamental aspects of laser theory. The simplicity of its evolution equations, equations (1.58)–(1.60), makes the model attractive. Its equivalence with the Lorenz equations [1], which have become the generic model to study chaos in ordinary differential equations, increases the relevance of the ring laser model. Of importance is the fact that the laser model lends itself quite naturally to a complexification of the variables. It suffices that the detuning be nonzero to have a coupling between the phase and the amplitude of the electric field and of the atomic polarization. This opens the door to an even richer phenomenology of complex behaviors.

The ring configuration for a laser is not simply an idealization intended for theoreticians. A number of lasers operate in this configuration. Dye lasers and some coherently pumped lasers are built with ring cavities. Laser gyroscopes are essentially ring lasers. If the ring cavity is perfectly symmetric with respect to the two directions of propagation, there is no preferential direction of oscillation and both directions must be taken into account. This is the simplest example of a multimode laser and we analyze some of its properties in Section 9.3. However, most lasers utilize Fabry–Pérot cavities and are perfectly stable. For this reason, they have attracted much less interest from the community of fundamental physicists. The other reason for avoiding the Fabry–Pérot laser equations is their complexity. It is true that a number of essential features are common to both the ring and the Fabry–Pérot cavities. However, the differences between these two configurations justify that we discuss in some detail their properties.

7.1 The semiclassical laser equations

Our starting point is the equations (1.18)–(1.20)

$$c^2 E_{zz} - E_{tt} = (N\mu/\varepsilon_0)(\mathcal{P}_{tt} + \mathcal{P}_{tt}^*) \tag{7.1}$$

$$\mathcal{P}_t = -i\omega_a \mathcal{P} - (i\mu/\hbar)E\mathcal{D} - \gamma_\perp \mathcal{P} \tag{7.2}$$

$$\mathcal{D}_t = -(2i\mu/\hbar)E(\mathcal{P} - \mathcal{P}^*) - \gamma_\parallel(\mathcal{D} - \mathcal{D}_a) \tag{7.3}$$

where we have used the notation $f_t \equiv \partial f/\partial t$. As a first step, we introduce a set of orthogonal functions $\phi(q, z)$ that satisfy the empty cavity equation $\phi_{zz}(q, z) + q^2\phi(q, z) = 0$. The orthogonality is reflected in the relation

$$\frac{1}{L}\int_0^L \phi^*(q, z)\phi(p, z)\, dz = \delta_{pq} \tag{7.4}$$

where δ_{pq} is the Kronecker symbol that equals unity if $p = q$ and zero otherwise. For a ring cavity, we use the unidirectional running waves $\phi(q, z) = \exp(iqz)$, whereas for the Fabry–Pérot cavity we use the standing waves $\phi(q, z) = 2^{1/2}\sin(qz)$. The periodic boundary condition for the ring cavity is $\phi(q, 0) = \phi(q, L)$, which gives $q = \pm 2\pi n/L$. For the Fabry–Pérot cavity, the boundary $\phi(q, 0) = \phi(q, L)$ gives $q = \pm n\pi/L$. In both cases we restrict q to positive values. For a typical laser cavity of about 1 meter long operating in the visible, $n = qL/2\pi$ is a large number of the order of 10^6. The decomposition of the electric field E on this basis is

$$E(z, t) = \sum_p \left(\phi(p, z)E(p, t) + c.c.\right) \tag{7.5}$$

that introduces the modal field amplitudes $E(p, t)$. Using the orthogonality relation (7.4) leads from Maxwell's equation (7.1) to

$$c^2 p^2 E(p, t) + E_{tt}(p, t) = -(N\mu/L\varepsilon_0)\int_0^L [\mathcal{P}_{tt}(z, t) + \mathcal{P}_{tt}^*(z, t)]\phi^*(p, z)\, dz \tag{7.6}$$

The next step is to separate in the modal field amplitude the fast oscillation from the slower time-dependence that results from the light–matter coupling. For this purpose we introduce the decompositions

$$E(p, t) = \mathcal{E}(p, t)\exp(-i\omega_p t)$$

$$\mathcal{P}(z, t) = (i\mu/\hbar)\sum_p \phi(p, z)\mathcal{P}(p, z, t)\exp(-i\omega_p t) \tag{7.7}$$

where $\omega_p = pc$. The prefactor $i\mu/\hbar$ is introduced to simplify the notation. The amplitudes $\mathcal{E}(p, t)$ and $\mathcal{P}(p, z, t)$ are complex functions. In view of the decomposition (7.5) with field amplitudes $E(p, t)$ that are space-independent, the decomposition of $\mathcal{P}(z, t)$ in terms of modal polarization amplitudes $\mathcal{P}(p, z, t)$ that still depend on space may seem contradictory. It is perfectly legitimate to expand both $\mathcal{P}(z, t)$ and the population inversion $\mathcal{D}(z, t)$ as the electric field, namely

$$\mathcal{P}(z, t) = \sum_p \phi(p, z)\mathcal{P}(p, t), \qquad \mathcal{D}(z, t) = \sum_p (\phi(p, z)\mathcal{D}(p, t) + c.c.) \tag{7.8}$$

The coupled evolution equations for the amplitude functions $\{\mathcal{E}(p, t), \mathcal{P}(p, t), \mathcal{D}(p, t)\}$ are especially intricate [2, 3]. They are useful for a perturbative expansion in powers of the electric dipole moment matrix element μ up to order three (only odd powers appear in this expansion for the field amplitudes), which gives the first nonlinear correction to the linear theory. This raises the additional difficulty that such an expansion requires a small parameter to converge, namely $(\mu/\hbar)^2 \ll 1$, whereas the single-mode lasing condition $A > 1$ derived in Chapter 1 implies $(\mu/\hbar)^2 > 2\gamma_\perp \varepsilon_0 \gamma_c/(\mathcal{D}_a N\hbar\omega_c)$, where γ_c is the field decay rate. This further reduces the domain of convergence of the perturbation expansion in powers of $\mu/\hbar$. Moreover, these equations have been derived and studied only in the good cavity limit where all atomic amplitudes $\{\mathcal{P}(p, t), \mathcal{D}(p, t)\}$ relax much faster than the field amplitudes and are therefore adiabatically eliminated. A necessary condition for this approximation to hold (away from critical points) is

$$\gamma_{cp} \ll \gamma_\perp, \qquad \gamma_\perp/\gamma_\parallel = \mathcal{O}(1) \tag{7.9}$$

where γ_{cp} is the cavity decay rate of mode p. However, most solid-state lasers in Fabry–Pérot cavities and some gas lasers as well belong to a different class of lasers, for which the inequalities among the decay rates are

$$\gamma_\parallel \ll \gamma_{cp} \ll \gamma_\perp \tag{7.10}$$

In this limit, there are two small parameters and therefore two time scales. It will appear later in this chapter [see, for instance, the result (7.50)] that the double limit $\gamma_\parallel/\gamma_{cp} \to 0$ and $\gamma_{cp}/\gamma_\perp \to 0$ is singular. We do not use the modal expansions (7.8) for these reasons. Instead, a mixed formulation in terms of the functions $\mathcal{E}(p, t)$, $\mathcal{P}(p, z, t)$, and $\mathcal{D}(z, t)$ leads to a more transparent derivation of the multimode equations.

With the expansions (7.7), the modal field equation becomes

$$\mathcal{E}_{tt}(p, t) - 2i\omega_p \mathcal{E}_t(p, t) = -\frac{iN\mu^2}{L\hbar\varepsilon_0} \int_0^L \phi^*(p, z) \sum_q (\phi(q, z)[\mathcal{P}_{tt}(q, z, t)$$
$$- 2i\omega_q \mathcal{P}_t(q, z, t) - \omega_q^2 \mathcal{P}(q, z, t)]e^{i(\omega_p - \omega_q)t}$$
$$- \phi^*(q, z)[\mathcal{P}_{tt}^*(q, z, t) + 2i\omega_q \mathcal{P}_t^*(q, z, t)$$
$$- \omega_q^2 \mathcal{P}^*(q, z, t)]e^{i(\omega_p + \omega_q)t}) dz \tag{7.11}$$

The integral on the right-hand side of this equation can be simplified considerably if we notice that it contains three different contributions:

1. The modal amplitude $\mathcal{P}(p, z, t)$ and its time derivatives, which are not affected by an oscillating exponential.

2. The modal amplitudes $\mathcal{P}(q, z, t)$, and their derivatives, with $p \neq q$ and $\omega_p \neq \omega_q$, which are affected by an exponential oscillating at the frequency difference $\omega_q - \omega_p$. These terms oscillate fast compared to the $\mathcal{P}(p, z, t)$ terms and may therefore be neglected.

3. The terms involving $\mathcal{P}^*(q, z, t)$ that oscillate at the sum frequency $\omega_q + \omega_p$ and are even more negligible than the previous terms.

An additional approximation is the slowly varying envelope approximation in the form introduced and justified in Section (1.3)

$$\omega_p|\mathcal{E}(p, t)| \gg |\partial\mathcal{E}(p, t)/\partial t|, \quad \omega_p|\mathcal{P}(p, z, t)| \gg |\partial\mathcal{P}(p, z, t)/\partial t| \quad (7.12)$$

This leaves us with the following field equations for the multimode Fabry–Pérot laser.

$$\mathcal{E}_t(p, t) = -\gamma_{cp}\mathcal{E}(p, t) - \frac{N\mu^2\omega_p}{2\hbar\varepsilon_0 L}\int_0^L |\phi(p, z)|^2\mathcal{P}(p, z, t)\,dz \quad (7.13)$$

We have introduced the phenomenological field damping rate or inverse photon lifetime γ_{cp} for each mode to account for the cavity losses.

The Bloch equations are derived from equations (7.2) and (7.3), using the expansions (7.7). For the polarization equation, we obtain

$$0 = \sum_p \big(\mathcal{P}_t(p, z, t) + [\gamma_\perp + i(\omega_a - \omega_p)]\mathcal{P}(p, z, t)$$
$$+ \mathcal{D}(z, t)[\mathcal{E}(p, t) + \mathcal{E}^*(p, t)e^{2i\omega_p t}]\big)\phi(p, z)e^{-i\omega_p t}$$
$$\cong \sum_p \big(\mathcal{P}_t(p, z, t) + [\gamma_\perp + i(\omega_a - \omega_p)]\mathcal{P}(p, z, t)$$
$$+ \mathcal{D}(z, t)\mathcal{E}(p, t)\big)\phi(p, z)e^{-i\omega_p t}$$
$$\equiv \sum_p \overline{\mathcal{P}}(p, z, t)\phi(p, z)e^{-i\omega_p t} \quad (7.14)$$

This last relation can be written as

$$\overline{\mathcal{P}}(p, z, t)\phi(p, z) + \sum_{q \neq p} \overline{\mathcal{P}}(q, z, t)\phi(q, z)e^{i(\omega_p - \omega_q)t} = 0 \quad (7.15)$$

As in the derivation of the field amplitude equation, we neglect terms that are multiplied by oscillations at the difference between two frequencies compared with terms that are not affected by an oscillating exponential. This leads to $\overline{\mathcal{P}}(p, z, t) = 0$, that is,

$$\mathcal{P}_t(p, z, t) = -[\gamma_\perp + i(\omega_a - \omega_p)]\mathcal{P}(p, z, t) - \mathcal{D}(z, t)\mathcal{E}(p, t). \quad (7.16)$$

For the population inversion, equation (7.3) yields

$$\begin{aligned}
\mathcal{D}_t(z, t) = {}&-\gamma_\|[\mathcal{D}(z, t) - \mathcal{D}_a] \\
&+ \frac{2\mu^2}{\hbar^2}\sum_{p,q}\Big(\phi(q, z)\phi^*(p, z)\mathcal{E}(q, t)\mathcal{P}^*(p, z, t)e^{i(\omega_p - \omega_q)t} \\
&+ \phi(q, z)\phi(p, z)\mathcal{E}(q, t)\mathcal{P}(p, z, t)e^{-i(\omega_p + \omega_q)t} + c.c.\Big) \\
\cong {}&-\gamma_\|[\mathcal{D}(z, t) - \mathcal{D}_a] + \frac{2\mu^2}{\hbar^2}\sum_p\big(|\phi(p, z)|^2\mathcal{E}(p, t)\mathcal{P}^*(p, z, t) + c.c.\big)
\end{aligned}$$

$$(7.17)$$

We have used the same approximation as in the derivation of the electric field and atomic polarization equations: amplitudes multiplied by fast oscillating terms are neglected compared to amplitudes that are not affected by an oscillating exponential. Eventually the basic equations for the multimode laser take the form

$$\mathcal{E}_t(p, t) = -\gamma_{cp}\mathcal{E}(p, t) - \frac{N\mu^2\omega_p}{2\hbar\varepsilon_0 L}\int_0^L |\phi(p, z)|^2\mathcal{P}(p, z, t)\,dz \quad (7.18)$$

$$\mathcal{P}_t(p, z, t) = -[\gamma_\perp + i(\omega_a - \omega_p)]\mathcal{P}(p, z, t) - \mathcal{D}(z, t)\mathcal{E}(p, t) \quad (7.19)$$

$$\mathcal{D}_t(z, t) = -\gamma_\|[\mathcal{D}(z, t) - \mathcal{D}_a] + \frac{2\mu^2}{\hbar^2}\sum_p |\phi(p, z)|^2\big(\mathcal{E}(p, t)\mathcal{P}^*(p, z, t) + c.c.\big)$$

$$(7.20)$$

These equations have been derived under the assumption that there is no degeneracy: To each p there is one and only one frequency ω_p. Therefore they are suitable for the study of Fabry–Pérot lasers and for unidirectional ring lasers. The main simplification introduced in the derivation of the basic equations (7.18)–(7.20) is the neglect of terms that oscillate in time. This implies that none of the frequency differences $\omega_p - \omega_q$ can become arbitrarily small. In fact, it implies more than this simple restriction. Had we used the modal expansions (7.8) for the atomic polarization and for the population inversion, we would have found that at order $(\mu/\hbar)^n$ contributions to the evolution of the electric field are affected by oscillating exponentials involving the sum of $n - 1$ frequency differences. Because cavity modes are equally spaced, the frequency differences between adjacent modes are equal and sums of frequency differences may vanish. In that case, the phenomenon of phase locking may appear over a range of laser parameters, in which the frequency differences are fixed though the whole spectrum may drift [2]. This requires an active phase–amplitude coupling, a property that has already disappeared in (7.18)–(7.20). Outside of the phase locking domain, the laser is called *free-running*. This chapter deals only with free-running lasers.

7.2 The rate equations

In this section we use the inequalities (7.10) among the three decay rates, $\gamma_\| \ll \gamma_{cp} \ll \gamma_\perp$, to simplify the evolution equations. These inequalities imply that we can adiabatically eliminate the atomic polarization. The formal procedure is to introduce a slow time $\tau = \gamma_\| t$ and expand the modal amplitudes $\{\mathcal{E}(p, t), \mathcal{P}(p, z, t), \mathcal{D}(z, t)\}$ in a power series of the small parameter $\varepsilon = \gamma_\|/\gamma_\perp$ as

$$\mathcal{E}(p, t) = (\hbar/\mu) \sqrt{\gamma_\| \gamma_\perp}\, \mathcal{E}(p, \tau, \varepsilon) = (\hbar/\mu) \sqrt{\gamma_\| \gamma_\perp}\, [\mathcal{E}_0(p, \tau) + \mathcal{O}(\varepsilon)]$$

$$\mathcal{P}(p, t) = (\hbar/\mu) \sqrt{\gamma_\|/\gamma_\perp}\, \mathcal{P}(p, \tau, \varepsilon) = (\hbar/\mu) \sqrt{\gamma_\|/\gamma_\perp}\, [\mathcal{P}_0(p, \tau) + \mathcal{O}(\varepsilon)]$$

$$\mathcal{D}(p, t) = \mathcal{D}(p, \tau, \varepsilon) = \mathcal{D}_0(p, \tau) + \mathcal{O}(\varepsilon) \tag{7.21}$$

With $\gamma_\|/\gamma_{cp} \gg \mathcal{O}(\varepsilon)$, all coefficients of the evolution equations remain finite in the limit $\gamma_\|/\gamma_\perp \to 0$. We then obtain $\mathcal{P}_0 = -\mathcal{D}_0 \mathcal{E}_0/[1 + i\delta(p)]$ with the dimensionless detuning being defined as $\delta(p) = (\omega_a - \omega_p)/\gamma_\perp$ and two coupled differential equations for $\mathcal{D}_0$ and $\mathcal{E}_0$. Expressing the relation for $\mathcal{P}_0$ back in the original variables yields

$$\mathcal{P}(p, z, t) = -\frac{\mathcal{D}(z, t)\mathcal{E}(p, t)}{\gamma_\perp [1 + i\delta(p)]} \tag{7.22}$$

We stress the fact that this relation is valid only to leading order in ε. Inserting this result into the modal field equation leads to

$$\mathcal{E}_t(p, t) = \mathcal{E}(p, t)\left(-\gamma_{cp} + \frac{N\mu^2 \omega_p}{\gamma_\perp [1 + i\delta(p)] 2\hbar\varepsilon_0 L} \int_0^L |\phi(p, z)|^2 \mathcal{D}(z, t)\, dz\right) \tag{7.23}$$

At this point we decompose the modal field amplitude into a real amplitude and a phase: $\mathcal{E}(p, t) \equiv |\mathcal{E}(p, t)| \exp[-i\Psi(p, t)]$. They are solutions of the equations

$$d|\mathcal{E}(p, t)|^2/dt =$$

$$|\mathcal{E}(p, t)|^2 \left(-2\gamma_{cp} + \frac{N\mu^2 \omega_p}{\gamma_\perp [1 + \delta^2(p)]\hbar\varepsilon_0 L} \int_0^L |\phi(p, z)|^2 \mathcal{D}(z, t)\, dz\right) \tag{7.24}$$

$$\Psi_t(p, t) = \frac{\delta_p N\mu^2 \omega_p}{2\gamma_\perp [1 + \delta^2(p)]\hbar\varepsilon_0 L} \int_0^L |\phi(p, z)|^2 \mathcal{D}(z, t)\, dz \tag{7.25}$$

Using the expression (7.22) for the polarization, the evolution equation of the population inversion becomes

$$\mathcal{D}_t(z, t) = -\gamma_\|[\mathcal{D}(z, t) - \mathcal{D}_a] - \frac{4\mu^2}{\hbar^2\gamma_\perp}\mathcal{D}(z, t)\sum_p \frac{|\phi(p, z)\mathcal{E}(p, t)|^2}{1 + \delta^2(p)} \quad (7.26)$$

The set of equations (7.24)–(7.26) forms the basis for the rate equation formulation of free-running multimode lasers. They have a rather special structure that has been the source of some misconceptions. It is apparent that the field phases $\Psi(p, t)$ play a special role in the laser dynamics. They are *passive* variables in the sense that they depend on the $N + 1$ dynamical variables (if there are N modes), which are the mode intensities and the population inversion. On the contrary, these $N + 1$ dynamical variables evolve independently of the phases and determine solely the stability of the system: They are the *active* variables. In many simplified treatments of the rate equations, the phases have been completely neglected and there are often statements that the phases are not defined in the rate equation limit. This misconception leads to serious difficulties when extensions of the rate equations are sought for describing more general situations. The best example is the case in which an external device (such as a mirror) is used to feed back part of the input field into the laser cavity. In that case, the injected field may act as a source term in the *field* equation (7.6) and not in the *intensity* equation (7.24). The result is a strong phase coupling and, in some cases, a complete reversal of the situation because the phase difference between the injected and the cavity fields may become the active variable.

7.2.1 The ring cavity rate equations

For a ring cavity, the $\phi(p, z)$ are running waves and their modulus is unity. Hence, if we define the space average of the population inversion

$$\mathcal{D}(t) = \frac{1}{L}\int_0^L \mathcal{D}(z, t)\, dz \quad (7.27)$$

the rate equations become

$$d|\mathcal{E}(p, t)|^2/dt = |\mathcal{E}(p, t)|^2\left(-2\gamma_{cp} + \frac{N\mu^2\omega_p\mathcal{D}(t)}{\gamma_\perp[1 + \delta^2(p)]\hbar\varepsilon_0}\right) \quad (7.28)$$

$$\Psi_t(p, t) = \frac{\delta_p N\mu^2\omega_p\mathcal{D}(t)}{2\gamma_\perp[1 + \delta^2(p)]\hbar\varepsilon_0} \quad (7.29)$$

$$\mathcal{D}_t(t) = -\gamma_\|[\mathcal{D}(t) - \mathcal{D}_a] - \frac{4\mu^2}{\hbar^2\gamma_\perp}\mathcal{D}(t)\sum_p \frac{|\mathcal{E}(p, t)|^2}{1 + \delta^2(p)} \quad (7.30)$$

Should the population inversion $\mathcal{D}_a$ in (7.3) depend on the spatial coordinate z, it would be replaced in the rate equation (7.30) by its space average

$$\mathcal{D}_a(z) \rightarrow \mathcal{D}_a = \frac{1}{L} \int_0^L \mathcal{D}_a(z) \, dz \qquad (7.31)$$

The main feature of these equations is that the N modal intensities $|\mathcal{E}(p, t)|^2$ and the average population inversion form a closed set of coupled evolution equations. Their main weakness is that they do not support multimode steady state solutions. For each mode, there are two solutions: The intensity $|\mathcal{E}(p, t)|^2$ either vanishes or has a finite value. A linear stability analysis of these steady state solutions indicates that the only stable solution corresponds to the mode with the highest gain having a finite intensity, whereas all other modes have zero intensity. Stable multimode operation can be restored, however. For instance, in a gas laser with a Doppler velocity distribution, the rate equations have additional couplings that lead to a stable multimode regime. However, in that case the coupling is no longer between the field intensities but between the field amplitudes.

7.2.2 The Fabry–Pérot cavity rate equations

The formulation of the rate equations for a Fabry–Pérot cavity is a difficult problem. In this section, we derive the Tang, Statz, and deMars equations [4], that are the classic rate equations for multimode solid-state lasers.

For a Fabry–Pérot cavity, the eigenfunctions used to expand the electric field are $\phi(p, z) = \sqrt{2} \sin(pz)$. Hence $|\phi(p, z)|^2 = 1 - \cos(2pz)$, and $|\phi(p, z)|^2 \cos(2qz) = \cos(2qz) - \frac{1}{2}\{\cos[2(p + q)z] + \cos[2(p - q)z]\}$. The modal intensity equation becomes

$$d|\mathcal{E}(p, t)|^2/dt = |\mathcal{E}(p, t)|^2 \left(-2\gamma_{cp} + \frac{N\mu^2 \omega_p}{\gamma_\perp [1 + \delta^2(p)]\hbar\varepsilon_0} [\mathcal{D}(t) - \mathcal{D}(p, t)] \right)$$

$$(7.32)$$

where we have defined the first moment of the population inversion[1]

$$\mathcal{D}(p, t) \equiv \frac{1}{L} \int_0^L \mathcal{D}(z, t) \cos(2pz) \, dz \qquad (7.33)$$

The main difference between the intensity equation for the ring and for the Fabry–Pérot lasers is the coupling with the population inversion grating through $\mathcal{D}(p, t)$. This coupling opens the door to many phenomena, the first of which is a multimode steady state. However, this coupling poses a problem because it leads to an infinite hierarchy of coupled equations. This can be seen from the

---

[1] Sometimes $\mathcal{D}(p, t)$ is defined with an additional factor 2 in front of the integral. We do not adopt this convention because it simplifies the formulation of the rate equations.

evolution equation for the space average of the population inversion $\mathcal{D}(t)$

$$\mathcal{D}_t(t) = \gamma_\parallel [\mathcal{D}_a - \mathcal{D}(t)] - \frac{4\mu^2}{\hbar^2 \gamma_\perp} \sum_p \frac{|\mathcal{E}(p, t)|^2}{1 + \delta^2(p)} [\mathcal{D}(t) - \mathcal{D}(p, t)] \quad (7.34)$$

and from the evolution equation for the general moment $\mathcal{D}(p, t)$, which is

$$\mathcal{D}_t(p, t) = -\left(\gamma_\parallel + \frac{4\mu^2}{\hbar^2 \gamma_\perp} \sum_{p'} \frac{|\mathcal{E}(p', t)|^2}{1 + \delta^2(p')} \right) \mathcal{D}(p, t)$$

$$+ \frac{2\mu^2}{\hbar^2 \gamma_\perp} \sum_{p'} \frac{|\mathcal{E}(p', t)|^2}{1 + \delta^2(p')} \left(\mathcal{D}(p + p', t) + \mathcal{D}(p - p', t) \right)$$

$$= -\left(\gamma_\parallel + \frac{4\mu^2}{\hbar^2 \gamma_\perp} \sum_{p'} \frac{|\mathcal{E}(p', t)|^2}{1 + \delta^2(p')} \right) \mathcal{D}(p, t) + \frac{2\mu^2}{\hbar^2 \gamma_\perp} \frac{|\mathcal{E}(p, t)|^2}{1 + \delta^2(p)} \mathcal{D}(t)$$

$$+ \frac{2\mu^2}{\hbar^2 \gamma_\perp} \sum_{p' \neq p} \frac{|\mathcal{E}(p', t)|^2}{1 + \delta^2(p')} \mathcal{D}(p - p', t)$$

$$+ \frac{2\mu^2}{\hbar^2 \gamma_\perp} \sum_{p'} \frac{|\mathcal{E}(p', t)|^2}{1 + \delta^2(p')} \mathcal{D}(p + p', t) \quad (7.35)$$

It is clear from the modal field amplitude equation (7.32) that both moments $\mathcal{D}(t)$ and $\mathcal{D}(p, t)$ are necessary to couple the field to the population grating and to allow for a multimode steady state. The Tang, Statz, and deMars approximation amounts to neglect all other moments, that is, the last two sums in (7.35). This leads to

$$\mathcal{D}_t(p, t) = -\left(\gamma_\parallel + \frac{4\mu^2}{\hbar^2 \gamma_\perp} \sum_{p'} \frac{|\mathcal{E}(p', t)|^2}{1 + \delta^2(p')} \right) \mathcal{D}(p, t) + \frac{2\mu^2}{\hbar^2 \gamma_\perp} \frac{|\mathcal{E}(p, t)|^2}{1 + \delta^2(p)} \mathcal{D}(t)$$

$$(7.36)$$

As any truncation in a moment expansion, it is especially difficult to justify this approximation. The success of the resulting theory – which has been used for thirty years to explain many facets of solid-state laser physics – is not in itself a proof. However, its perverse effect is to demotivate a search for a justification of these rate equations. The loss is that we do not know the domain of validity of the Tang, Statz, and deMars equations.

Finally, we introduce a set of reduced variables

$$I(p, \tau) = \frac{4\mu^2}{\hbar^2 \gamma_\perp \gamma_\parallel} \frac{|\mathcal{E}(p, t)|^2}{1 + \delta^2(1)}, \qquad D(\tau) = \frac{N\mu^2 \omega_1}{2\hbar\varepsilon_0 \gamma_\perp \gamma_{c1}} \frac{\mathcal{D}(t)}{1 + \delta^2(1)}$$

$$D(p, \tau) = \frac{N\mu^2 \omega_1}{2\hbar\varepsilon_0 \gamma_\perp \gamma_{c1}} \frac{\mathcal{D}(p, t)}{1 + \delta^2(1)}, \qquad w = \frac{N\mu^2 \omega_1 \mathcal{D}_a}{2\hbar\varepsilon_0 \gamma_\perp \gamma_{c1} [1 + \delta^2(1)]}$$

$$\tau = \gamma_\| t, \quad k_p = 2\gamma_{cp}/\gamma_\|, \quad \Gamma_p = \frac{1 + \delta^2(1)}{1 + \delta^2(p)}, \quad \gamma_p = \frac{\gamma_{c1}[1 + \delta^2(1)]\,\omega_p}{\gamma_{cp}[1 + \delta^2(p)]\,\omega_1}$$

$$(7.37)$$

In these definitions, w is the pump parameter. It is normalized so that the laser first threshold occurs at $w = 1$. It is equal to $\alpha_1/\{\gamma_{c1}[1 + \delta^2(1)]\}$, where α_1 is the linear gain of mode $p = 1$ on resonance and per unit length, as defined in (1.35). With these reduced variables, the Tang, Statz, and deMars rate equations are

$$I_\tau(p, \tau) = k_p\big(\gamma_p[D(\tau) - D(p, \tau)] - 1\big)I(p, \tau) \qquad (7.38)$$

$$\Psi_\tau(p, \tau) = k_p\delta(p)\gamma_p[D(\tau) - D(p, \tau)]/2 \qquad (7.39)$$

$$D_\tau(\tau) = w - \left(1 + \sum_p I(p, \tau)\Gamma_p\right)D(\tau) + \sum_p I(p, \tau)\Gamma_p D(p, \tau) \quad (7.40)$$

$$D_\tau(p, \tau) = -\left(1 + \sum_q I(q, \tau)\Gamma_q\right)D(p, \tau) + \frac{1}{2}I(p, \tau)\Gamma_p D(\tau) \quad (7.41)$$

As expected from our previous discussion, the phase $\Psi(p, t)$ of the modal field is well defined but is a passive variable. The remaining $2N + 1$ equations describe the dynamics of N modes in a Fabry–Pérot laser.

7.3 The Tang, Statz, and deMars equations

In this and the following sections, we analyze the Tang, Statz, and deMars (TSD) rate equations. For most lasers to which these equations apply, the N modes that oscillate simultaneously are spread over a small frequency range. It is therefore customary to introduce two simplifications related to this property. The first approximation is to assume equal loss rates for all modes: $k_p \equiv k$. The second simplification is to approximate the ratio ω_1/ω_p by unity. Hence $\Gamma_p = \gamma_p$ and the simplified TSD equations become

$$I_\tau(p, \tau) = k\big(\gamma_p[D(\tau) - D(p, \tau)] - 1\big)I(p, \tau) \qquad (7.42)$$

$$D_\tau(\tau) = w - \left(1 + \sum_p I(p, \tau)\gamma_p\right)D(\tau) + \sum_p I(p, \tau)\gamma_p D(p, \tau) \quad (7.43)$$

$$D_\tau(p, \tau) = -\left(1 + \sum_q I(q, \tau)\gamma_q\right)D(p, \tau) + \frac{1}{2}I(p, \tau)\gamma_p D(\tau) \quad (7.44)$$

The phase equation (7.39) is modified only by the replacement of k_p by k.

7.3.1 Steady state solutions

The steady state solution of the TSD equations is easily found by setting $I_\tau(p) = D_\tau(\tau) = D_\tau(p, \tau) = 0$. The first equation, (7.42), gives a relation between $D(p)$ and D:

$$D(p) = D - 1/\gamma_p \qquad (7.45)$$

Inserting this result into (7.44) and summing over p gives a relation between $\Sigma(1)$ and D, where $\Sigma(n) = \sum_p I(p)\gamma_p^n$. Dividing (7.44) by γ_p and then summing over p, we obtain a relation between $\Sigma(0)$ and D. Using these expressions for $\Sigma(0)$ and $\Sigma(1)$ and (7.45), we obtain from (7.43) a closed equation for D

$$w = D + \frac{DS_1 - S_2}{S_1 - (N - 1/2)D}, \qquad S_1 = \sum_p 1/\gamma_p, \qquad S_2 = \sum_p 1/\gamma_p^2$$

$$(7.46)$$

Finally, the modal intensities are expressed in terms of D and from (7.44) we get

$$I(p) = \frac{1}{\gamma_p} \frac{D - 1/\gamma_p}{S_1 - (N - 1/2)D} \qquad (7.47)$$

The neat feature that makes the simpler version (7.42)–(7.44) of the TSD equations so useful is the fact that the steady state properties are entirely determined by a quadratic equation for the average population, equation (7.46). This considerably simplifies the analysis of the problem and makes some numerical simulation much easier to perform since the steady state can be computed exactly for any mode number.

7.3.2 Stability of the single-mode solution

The single-mode steady state solution is

$$D = 2 + w/2 - \sqrt{2 + w^2/4}, \qquad I = -2 + w/2 + \sqrt{2 + w^2/4}$$

$$D(1) = D - 1 \qquad (7.48)$$

At threshold, defined by $w = 1$, we have $D = 1$, $I = D(1) = 0$. The next step is to perform a linear stability analysis of this steady state. Seeking solutions of the TSD equations in the form $Z(\tau) = Z + \varepsilon Z_1 \exp(\lambda\tau) + \mathcal{O}(\varepsilon^2)$, where Z is any of the three variables D, $D(1)$, and I, we obtain a characteristic equation

$$\lambda^3 + 2\lambda^2(1 + I) + \lambda[k(w - 1) + (1 + I)^2 - I^2/2] + kI(4 + I - D)/2 = 0$$

$$(7.49)$$

The limit $k \gg 1$ in which the solid state lasers operate can be used to simplify the expression of the three roots of the characteristic equation. These roots are

$$\lambda_1 = -\frac{2 + (D-2)^2}{(D-2)(D-4)} + \mathcal{O}(1/k)$$

$$\lambda_{2,3} = \pm i\sqrt{k(w-1)} - \frac{3[2 - (D-2)^2]}{2(D-2)(D-4)} + \mathcal{O}(k^{-1/2})$$

$$(7.50)$$

The real part of all three roots is negative and no instability of the steady state can happen. However, the presence of complex roots implies that a small perturbation of the steady state relaxes via damped oscillations that are characterized by the so-called relaxation oscillation frequency

$$\Omega_R^2 = k(w-1) \tag{7.51}$$

The remarkable property of this result is that this frequency is identical to the unidirectional ring laser relaxation oscillation frequency obtained in the same limit (7.10). Indeed, using the notation (7.37), the single-mode unidirectional ring laser rate equations (7.28) and (7.29) are

$$I_\tau = kI(D-1), \qquad D_\tau = w - D - ID \tag{7.52}$$

The finite intensity steady state solution is $I = w - 1$ and $D = 1$. Its linear stability yields the eigenvalues

$$\lambda = \frac{1}{2}\left(-w \pm \sqrt{w^2 - 4k(w-1)}\right)$$

$$= \pm i\Omega_R + \mathcal{O}(1) \tag{7.53}$$

in the large k limit. The same frequency is also found in Section 9.3.2, when we discuss the stability of the counterpropagating solutions in the ring laser. Note that we have

$$\Omega_R \tau = \sqrt{2\gamma_c \gamma_\parallel (w-1)} t \tag{7.54}$$

which gives the scaling of the frequency on the physical time scale. The factor 2 appearing in this expression of the relaxation oscillation may be a source of confusion. It appears only because we use γ_c, the damping rate of the field amplitude, instead of the intensity damping rate, which is equal to $2\gamma_c$.

Another aspect of the eigenvalues (7.50) is worthy of comment and is common to ring lasers as well when the inequalities (7.10) are satisfied. Since $k \gg 1$, the three roots define two time scales. The fast time scale is characterized by oscillations at the relaxation frequency (7.51) that are $\mathcal{O}(k^{1/2})$, whereas the damping rates, which are the real parts of the roots, are $\mathcal{O}(1)$. Thus oscillations and damping operate on very different time scales. This means that if we want to modify the nature of the laser output via a periodic modulation of a control parameter (typically, the gain or the losses) we need to select a modulation frequency that is $\mathcal{O}(k^{1/2})$ but an amplitude that is only $\mathcal{O}(1)$ to have a competition between the internal relaxation and the external perturbation.

References

[1]   H. Haken, *Phys. Lett.* **A53** (1975) 77.
[2]   W. E. Lamb, Jr., *Phys. Rev.* **134** (1964) A1429.
[3]   M. Sargent III, M. O. Scully, and W. E. Lamb, Jr., *Laser Physics* (Addison-Wesley, Reading, 1974).
[4]   C. L. Tang, H. Statz, and G. deMars, *J. Appl. Phys.* **34** 2289 (1963).

8

Antiphase dynamics

In a recent series of experimental papers, it has appeared that lasers that are well described by the simplified TSD rate equations (7.42)–(7.44) display what is called, in laser physics, *antiphased dynamics*. Its simplest manifestation is that when N modes oscillate, the total intensity displays many properties that are those of a single-mode laser. In the simplest experiment, the laser is initially in a steady state. A control parameter is suddenly changed and the relaxation toward the new steady state is recorded. This transient evolution is then Fourier analyzed to evaluate its frequency content. The result of this experiment is that each mode is characterized by as many frequencies as there are modes, whereas the total intensity, which is simply the sum of all modal intensities, is characterized by only one frequency. This frequency is the single-mode relaxation oscillation frequency (7.51) and is also the highest of the N frequencies. This property has been observed with a Nd-doped optical fiber laser [1]. Using a LiNdP$_4$O$_{12}$ (abbreviated as *LNP*) laser oscillating on two or three modes, it was shown that the noise spectrum of the laser displayed the same antiphase dynamics [2]. With the same laser, antiphase dynamics was also reported in the case where a feedback loop induces a chaotic output [3]. The feedback loop has the effect of injecting part of the output beam in the laser after each modal intensity has been subjected to a modulation. Antiphase dynamics also plays a significant role when a multimode laser undergoes a Feigenbaum cascade toward chaos. It has been reported that though the total intensity displays this usual route to chaos, the modal intensities do not [4]. Another aspect of antiphase dynamics appears when the system is in a self-pulsing regime. The best documented laser to display this transition to periodic states is the laser with intracavity second harmonic generation. Antiphase dynamics in this laser has been reported experimentally [5] and a number of theoretical results have been obtained as well [6]–[11]. In this chapter we deal with only antiphase dynamics in the TSD rate equations.

8.1 The reference model

From analytical and numerical studies of the TSD equations [12]–[14], it has been concluded that two key factors for the occurrence of antiphase dynamics are the large value of k and relative gains γ_p close to unity. This suggests that we investigate a simple model that is built as follows. We introduce a small parameter

$$\varepsilon \equiv 1/\sqrt{k} \ll 1 \tag{8.1}$$

in terms of which we define new variables that are the deviations from the steady state

$$D(\tau) = D + \varepsilon n(\mathrm{T}, \varepsilon), \qquad D(p, \tau) = D(p) + \varepsilon n(p, \mathrm{T}, \varepsilon)$$

$$I(p, \tau) = I(p) + s(p, \mathrm{T}, \varepsilon) \tag{8.2}$$

where $\mathrm{T} = \tau/\varepsilon$ is the new scaled time. It is assumed that the deviations have an expansion $f(\varepsilon) = f_0 + \varepsilon f_1 + \mathcal{O}(\varepsilon^2)$, where $f = \{n(\mathrm{T}, \varepsilon), n(p, \mathrm{T}, \varepsilon), s(p, \mathrm{T}, \varepsilon)\}$. Because antiphase dynamics is most pronounced for γ_p close to unity, we express this restriction by expanding the relative gains in series of ε around unity.

$$\gamma_p = 1 - \varepsilon g(p, \varepsilon), \qquad g(p, \varepsilon) = \mathcal{O}(1) \tag{8.3}$$

This makes the steady state solutions (7.45)–(7.47) functions of ε. If we expand them in powers of ε, we find that to dominant order in ε we have $I(p) \equiv \mathcal{I}$ and $D(p) \equiv \mathcal{D}$ with

$$\mathcal{D} = D - 1, \qquad \mathcal{I} = (w - D)/N, \qquad w = D + \frac{N(D-1)}{N - (N - 1/2)D} \tag{8.4}$$

The deviations defined by (8.2) obey a set of $2N + 1$ equations obtained from (7.42)–(7.44). To leading order in ε, these equations are

$$dn/d\mathrm{T} = -\sum_q s(q), \qquad dn(p)/d\mathrm{T} = s(p)D/2 - D\sum_q s(q)$$

$$ds(p)/d\mathrm{T} = [\mathcal{I} + s(p)][n - n(p)] \tag{8.5}$$

At this level of description, k contributes only to the definition of the time scale T. As a first step toward the study of this problem, we consider the linearized equations

$$dn/d\mathrm{T} = -\sum_q s(q), \qquad dn(p)/d\mathrm{T} = s(p)D/2 - D\sum_q s(q)$$

$$ds(p)/d\mathrm{T} = \mathcal{I}[n - n(p)] \tag{8.6}$$

From these equations we find that the evolution equation for the total intensity I is given by a closed equation

$$d^2 I/d\tau^2 + (w - 1)I = 0, \qquad I = \sum_p s(p, \tau) \tag{8.7}$$

In this approximation, the total intensity oscillates with a single frequency that is the single-mode relaxation oscillation frequency (7.51). The same is true for the population average $n(\tau)$.

Since the asymptotic model (8.6) is linear with constant coefficients, it can be solved exactly. To determine the eigenvalues, the simplest way is to express the system (8.6) in the matrix form $dz/d\tau = \mathcal{M}z$ for the column vector $z = \{s(1), n(1), s(2), n(2), \ldots, s(N-1), n(N-1), I, \Sigma n, n\}$, where Σn is the sum over all $n(p)$, and to solve the eigenvalue problem $\mathcal{M}z = \lambda z$. In this representation, the characteristic equation $\det(\mathcal{M} - \lambda 1) = 0$, where 1 is the unit matrix, becomes

$$\begin{vmatrix} M_1 & 0 & \cdots & 0 & 0 & Q \\ 0 & M_2 & \cdots & 0 & 0 & Q \\ \cdots & \cdots & \cdots & \cdots & & \cdots \\ \cdots & \cdots & \cdots & \cdots & & \cdots \\ 0 & 0 & \cdots & M_{N-1} & & Q \\ 0' & 0' & \cdots & & 0' & M_N \end{vmatrix} = 0 \tag{8.8}$$

with

$$M_1 = M_2 = \cdots = M_{N-1} = \begin{vmatrix} \lambda & \mathcal{I} \\ -D/2 & \lambda \end{vmatrix}$$

$$M_N = \begin{vmatrix} \lambda & \mathcal{I} & -N\mathcal{I} \\ N\mathcal{D} - D/2 & \lambda & 0 \\ 1 & 0 & \lambda \end{vmatrix}$$

$$0 = \begin{vmatrix} 0 & 0 \\ 0 & 0 \end{vmatrix}, \qquad 0' = \begin{vmatrix} 0 & 0 \\ 0 & 0 \\ 0 & 0 \end{vmatrix} \qquad Q = \begin{vmatrix} 0 & 0 & -\mathcal{I} \\ \mathcal{D} & 0 & 0 \end{vmatrix} \tag{8.9}$$

The solution of (8.8) is $(M_1)^{N-1} M_N = 0$. From $M_1 = 0$, we obtain the roots

$$\lambda_L = \pm i\Omega_L, \qquad \Omega_L^2 = \mathcal{I}D/2 \tag{8.10}$$

From $M_N = 0$, we obtain the roots

$$\lambda_0 = 0, \qquad \lambda_R = \pm i\Omega_R,$$

$$\Omega_R^2 = \mathcal{I}(D/2 - DN + N) = \mathcal{I}N + D - 1 = w - 1 \qquad (8.11)$$

The low frequency Ω_L is degenerate and has $N - 1$ eigenvectors. Thus, in this approximation the multimode rate equations are characterized by only two frequencies. Note the relation between the two frequencies

$$\Omega_L^2 - \Omega_R^2 = \mathcal{I}N(D - 2) \qquad (8.12)$$

The eigenvectors have a simpler physical meaning in the representation $z = \{n, n(1), n(2), \ldots, (N), s(1), s(2), \ldots, s(N)\}$. Using the results (8.10) and (8.11) for the eigenvalues, we easily find the eigenvectors

$$\lambda_0 = 0, \qquad z_0 = \{n = 1, n(p) = 1, s(p) = 0; p = 1, 2, \ldots, N\}$$

$$\lambda_L = \pm i\Omega_L, \qquad z(q) = \{n = 0, n(1) = 1/2, n(r) = -\delta_{rq}/2$$

$$s(1) = \lambda_L/D, s(r) = -\delta_{rq}\lambda_L/D; r = 2, 3, \ldots, N\}$$

$$\lambda_R = \pm i\Omega_R, \qquad z_R = \{n = N, n(p) = \beta, s(p) = -\lambda_R; p = 1, 2, \ldots, N\}$$

$$(8.13)$$

where

$$q = 2, 3, \ldots, N \qquad \text{and} \qquad \beta = N(1 - D)/(w - D) \qquad (8.14)$$

The global variables (total intensity I and average population inversion n) depend only on Ω_R. The $2N$ modal variables [intensity $s(q)$ and population grating $n(q)$] depend on both frequencies.

Using the definition of $\mathcal{I}$ and equation (8.4) for D leads to

$$\Omega_R^2/\Omega_L^2 = 2(w - 1)/\mathcal{I}D$$

$$= \frac{2N}{D}\frac{w - 1}{w - D}$$

$$= 1 - 2N + 4N/D \qquad (8.15)$$

From this expression, two limits are easily analyzed:

1. $w = 1 + \eta, \qquad 0 < \eta \ll 1, \qquad N = \mathcal{O}(1)$

$$D = 1 + \eta/(2N + 1) + \mathcal{O}(\eta^2)$$

$$\mathcal{I} = 2\eta/(2N + 1) + \mathcal{O}(\eta^2)$$

$$\Omega_L^2 = \eta/(2N + 1) + \mathcal{O}(\eta^2)$$

$$\Omega_R^2 = \eta + \mathcal{O}(\eta^2)$$

$$\Omega_R^2/\Omega_L^2 = 2N + 1 + \mathcal{O}(\eta) \qquad (8.16)$$

2. $w \gg 1$, $N = \mathcal{O}(1)$

$$D = \frac{2N}{2N - 1}\left(1 - \frac{1}{w(2N - 1)}\right) + \mathcal{O}(1/w^2)$$

$$\mathcal{I} = w(2N - 1)/N + \mathcal{O}(1)$$

$$\Omega_L^2 = w + \mathcal{O}(1)$$

$$\Omega_R^2 = \mathcal{I}/N + \mathcal{O}(1)$$

$$\Omega_R^2/\Omega_L^2 = 2N - 1 + \mathcal{O}(1/w) \tag{8.17}$$

The remarkable property that appears from these relations is that the ratio Ω_R^2/Ω_L^2 varies from $2N + 1$ to $2N - 1$ because it can be shown that it is a monotonically decreasing function of w. For a two-mode laser, this ratio varies therefore between 3 and 5. From (8.4) and (8.15), we find that the special value $\Omega_R = 2\Omega_L$ is reached for $D = 8/7$, which implies the relatively low pump $15/7 \simeq 2.143$. For that value of the pump parameter, a strong coupling between the two frequencies is expected. As a result, a frequency-selective perturbation, such as a periodic modulation of some laser parameter (gain, loss, or length, for instance) applied to the laser at Ω_R (at Ω_L, respectively) is expected to induce a large amplitude response at Ω_L (at Ω_R, respectively) as well.

8.2 Time-dependent solution

An additional insight into the antiphase dynamics can be gained by considering the general solution of the linearized system (8.6). We define a state vector $S = \{n, n(p), s(p); p = 1, 2, \ldots, N\}$. Using the eigenvectors (8.13), we have

$$S(T) = C_0 z_0 + \sum_{p=2}^{N} \left(C_p z(p) \exp(i\Omega_L T) + c.c. \right) + \left(C_R z_R \exp(i\Omega_R T) + c.c. \right) \tag{8.18}$$

Inserting this expression in the linearized equations yields the solution

$$n(T) = n(0) - \frac{N\delta}{N - \beta} + \frac{N\delta}{N - \beta}\cos(\Omega_R T) - \frac{N\Delta}{\Omega_R}\sin(\Omega_R T) \tag{8.19}$$

$$n(p, T) = n(0) - \frac{N\delta}{N - \beta} + (\delta - \delta_p)\cos(\Omega_L T) + \frac{D}{2\Omega_L}(\Delta_p - \Delta)\sin(\Omega_L T)$$

$$+ \frac{\beta\delta}{N - \beta}\cos(\Omega_R T) - \frac{\beta\Delta}{\Omega_R}\sin(\Omega_R T) \tag{8.20}$$

$$s(p, \text{T}) = \frac{\Omega_L}{D}(\delta_p - \delta)\sin(\Omega_L\text{T}) + (\Delta_p - \Delta)\cos(\Omega_L\text{T})$$

$$+ \frac{\Omega_R\delta}{N - \beta}\sin(\Omega_R\text{T}) + \Delta\cos(\Omega_R\text{T}) \qquad (8.21)$$

in terms of

$$\delta_p = n(0) - n(p, 0), \qquad \Delta_p = s(p, 0)$$

$$\delta = \frac{1}{N}\sum_{q=1}^{N}\delta_q, \qquad \Delta = \frac{1}{N}\sum_{q=1}^{N}\Delta_q \qquad (8.22)$$

The parameter β has been defined in (8.14). From the expression for the modal intensities (8.21) it follows that the total intensity is given by

$$I(\text{T}) = \sum_{q=1}^{N}s(q, \text{T}) = N\left(\frac{\Omega_R\delta}{N - \beta}\sin(\Omega_R\text{T}) + \Delta\cos(\Omega_R\text{T})\right) \qquad (8.23)$$

Although these solutions are produced by the linearized approximation of the asymptotic model (8.5), they already provide a number of useful features.

- We recover the property that the global variables $n(\text{T})$ and $I(\text{T})$ oscillate only at the single-mode frequency Ω_R.
- Experimentally, the conventional way to characterize the frequency content of a signal is via its power spectrum. In the modal intensities $s(p, \text{T})$, the amplitude of the oscillations at the high frequency Ω_R does not depend on the modal index p. The corresponding peak height in the power spectrum, which is equal to the modulus squared of the coefficient of $\exp(i\Omega_R\text{T})$, is the same for all modal intensities. Let $P(x_q, \Omega)$ be the the peak height of the power spectrum of the variable $x(q, \text{T})$ at frequency Ω and $P(\Sigma x, \Omega)$ the peak height of the power spectrum of the sum of all $x(q, \text{T})$ at frequency Ω. The following relations are easily obtained from (8.21) and (8.23):

$$P(s_q, \Omega_R) = P(s, \Omega_R) \qquad (8.24)$$

$$P(\Sigma s, \Omega_R) = N^2 P(s, \Omega_R) \qquad (8.25)$$

$$P(\Sigma s, \Omega_L) = 0 \qquad (8.26)$$

These relations are independent of the initial condition, that is, of the system preparation because they relate peak heights *at the same frequency*, which means relations among the components of one of the eigenvectors (8.13). On the contrary, relations between peaks of the power spectra at different frequencies depend on relations between components of different eigenvectors and therefore, from (8.18), involve the preparation of the system through the coefficients C_0, C_p, and C_R.

- The peak at Ω_R in the power spectrum of the population inversion grating is also mode-independent: $P(n_q, \Omega_R) = P(n_p, \Omega_R)$.

8.3 Power spectrum identities

The occurrence of relations such as (8.25)–(8.26) is important because they are intrinsic properties of the power spectrum, irrespective of the preparation of the system. Furthermore, they do not depend on, and do not require a knowledge of the parameters of the laser. The relation (8.25) requires only two properties: linear evolution equations and equal peaks in the modal intensity power spectra, which is expressed by (8.24). This relation expresses a degeneracy due to the limit $\gamma_p = 1$. If the relative gain functions γ_p are different from unity, the equality (8.24) is no longer valid and the relation between the total and the modal intensity power spectra is more complex. The cancellation of $P(\Sigma s, \Omega_L)$ is eventually related to symmetry properties of the eigenvectors (8.13) and are also intimately related to the approximation $\gamma_p = 1$. For arbitrary values of the relative gain, it does not seem possible to generalize the relations (8.25)–(8.26) to an arbitrary number of modes. However, in the case of two modes, it is still possible to have useful explicit relations.

To derive these relations, let us first consider the linearized equations (8.6) for two modes and an arbitrary relative gain parameter $\gamma_2 \equiv \gamma$

$$dn/d\tau = -s(1) - s(2)$$

$$dn(1)/d\tau = s(1)D/2 - D(1)[s(1) + \gamma s(2)]$$

$$dn(2)/d\tau = \gamma s(2)D/2 - D(2)[s(1) + \gamma s(2)]$$

$$ds(1)/d\tau = I(1)[n - n(1)]$$

$$ds(2)/d\tau = I(2)\gamma[n - n(2)] \tag{8.27}$$

The solution of these linearized equations is

$$
\begin{pmatrix} n \\ n(1) \\ n(2) \\ s(1) \\ s(2) \end{pmatrix} = C_1 \begin{pmatrix} -1 - S_L \\ 1 - D/2 + (1 - D)\gamma S_L \\ \left(1 - \dfrac{\gamma D}{2}\right)S_L + \dfrac{1}{\gamma} - D \\ i\Omega_L \\ iS_L\Omega_L \end{pmatrix} e^{i\Omega_L\tau}
$$

$$+ C_2 \begin{pmatrix} -1 - S_R \\ 1 - D/2 + (1 - D)\gamma S_R \\ \left(1 - \frac{\gamma D}{2}\right)S_R + \frac{1}{\gamma} - D \\ i\Omega_R \\ iS_R\Omega_R \end{pmatrix} e^{i\Omega_R \tau} + c.c. + C_3 \begin{pmatrix} 1 \\ 1 \\ 1 \\ 0 \\ 0 \end{pmatrix} \qquad (8.28)$$

In the solution (8.28), the two frequencies are given by

$$\Omega_{L,R}^2 = \{2 - D/2 + [1 - \gamma(D - 1)]S_{L,R}\}I(1) \qquad (8.29)$$

and the coefficients S_L and S_R are defined through

$$S_L = S_-, \qquad S_R = S_+, \qquad S_\pm = \frac{(\gamma - 1)(4 - \gamma D^2) \pm Q}{4\gamma(D - 1)[1 - \gamma(D - 1)]}$$

$$Q^2 = (1 - \gamma)^2(4 - \gamma D^2)^2 + 16(D - 1)(\gamma D - 1)[1 - \gamma(D - 1)]^2 \qquad (8.30)$$

If $\gamma = 1, S_\pm = \pm 1$. For $\gamma < 1$, the inequalities $S_L < 0$ and $S_R > 0$ are always satisfied. An interesting relation between the averaged population inversion and the two modal intensities is

$$\gamma(D - 1)I(2) = (D - 1/\gamma)I(1) \qquad (8.31)$$

The solution (8.28) yields at once the identities $P(s_2, \Omega_J) = S_J^2 P(s_1, \Omega_J)$ and $P(s_1 + s_2, \Omega_J) = (1 + S_J)^2 P(s_1, \Omega_J)$, where $J = R$ or L. We write s_j for $s(j)$ to lighten the notation. From these relations we derive the generalization of the power spectra relations (8.25)–(8.26) for $\gamma < 1$

$$P(\Sigma s, \Omega_R) = \left(\sqrt{P(s_1, \Omega_R)} + \sqrt{P(s_2, \Omega_R)}\right)^2$$

$$P(\Sigma s, \Omega_L) = \left(\sqrt{P(s_1, \Omega_L)} - \sqrt{P(s_2, \Omega_L)}\right)^2 \qquad (8.32)$$

Conversely, the verification of these relations indicates that, in addition to the validity of the linearized theory, the system is characterized by two widely separated time scales: The internal frequencies dominate on the fast time scale, and damping acts on a much longer time scale.

To test the domain of validity of this relation, we have integrated numerically the complete rate equations (7.42)–(7.44). The initial condition is the steady state for $w = 2.5$. The time integration is performed with $w = 2.25$ (a 10% drop from the pump parameter of the initial condition) and $k = 10^4$, which is typical for Nd:YAG lasers. The results are given in Tables 8.1 and 8.2 for the two frequencies.

The precision on the calculated results is better than 0.001. The first three columns result from the numerical integration. The last column is obtained by

Table 8.1 *Peaks of the power spectra at* Ω_R *for* $k = 10^4$ *and* $w = 2.25$ *normalized to* $P(\Sigma s, \Omega_R)$. *The initial condition is the steady state for* $w = 2.5$.

γ	$\sqrt{P(s_1, \Omega_R)}$	$\sqrt{P(s_2, \Omega_R)}$	$P(\Sigma s, \Omega_R)$	$\left(\sqrt{P(s_1, \Omega_R)} + \sqrt{P(s_1, \Omega_R)}\right)^2$
0.999	0.502	0.498	1	1.000
0.99	0.520	0.480	1	1.000
0.95	0.604	0.396	1	1.000
0.90	0.714	0.286	1	1.000
0.85	0.821	0.179	1	1.000
0.80	0.921	0.079	1	1.000

Table 8.2 *Peaks of the power spectra at* Ω_L *for* $k = 10^4$ *and* $w = 2.25$ *normalized to* $P(\Sigma s, \Omega_R)$. *The initial condition is the steady state for* $w = 2.5$.

γ	$\sqrt{P(s_1, \Omega_L)}$	$\sqrt{P(s_2, \Omega_L)}$	$P(\Sigma s, \Omega_L)$	$\left(\sqrt{P(s_1, \Omega_L)} - \sqrt{P(s_2, \Omega_L)}\right)^2$
0.999	0.015	0.015	< 0.001	< 0.001
0.99	0.063	0.062	< 0.001	< 0.001
0.95	0.215	0.232	< 0.001	< 0.001
0.90	0.148	0.166	< 0.001	< 0.001
0.85	0.232	0.299	0.073	0.0045
0.80	0.289	0.385	0.119	0.0092

using the numerical results of the second and third columns. Although the sum and difference relations (8.32) rely on properties of the linearized equations, the sum relation (at Ω_R) is an excellent approximation for a large domain of the relative gain. The difference relation (at Ω_L) remains a good relation as long as the peak in the total intensity has vanishing amplitude; that is, as long as there is antiphase. For each value of the relative gain, all peaks have been divided by $P(\Sigma s, \Omega_R)$. A better agreement is obtained for a smaller variation between the pump parameter of the initial condition and of the time-dependent evolution. In that case the linear approximation is better justified.

8.4 How many frequencies?

The reference model indicates that in the double limit $k \gg 1$ and $\gamma_p = 1$, there are two frequencies Ω_L and Ω_R. The main limitation of the reference model comes from the absence of any decay rate corresponding to a real part of the roots λ obtained in (8.13). To assess the influence of these decay rates, we consider a slight generalization of the reference model defined in Section 8.1. We start from the TSD rate equations (7.42)–(7.44), where we set $\gamma_p = 1$ for all modes but we do not use the expansion in powers of $k^{-1/2}$. In this way we keep the cavity loss influence on the eigenvalues intact. Hence the steady state solutions are given by (8.4). A linear stability of these solutions gives the characteristic equation

$$[P(2, \lambda)]^{N-1} P(3, \lambda) = 0$$

$$P(2, \lambda) = 2N\lambda^2 + 2N\lambda(1 + w - D) + Dk(w - D) \tag{8.33}$$

$$
\begin{aligned}
P(3, \lambda) = 2N\lambda^3 &+ 4N\lambda^2[1 + w - D] + \lambda[2N - 4ND + (2N - 1)D^2 \\
&- 4NDk + (2N - 1)D^2 k + 4Nw - (4N - 2)Dw + 4Nkw \\
&- (2N - 1)Dkw + (2N - 1)w^2] - 4NDk + (4N - 2)D^2 k \\
&+ 4Nkw - 3(2N - 1)Dkw + (2N - 1)kw^2
\end{aligned}
\tag{8.34}
$$

8.4.1 Large mode number limit: One frequency

The quadratic equation $P(2, \lambda) = 0$ has complex conjugate roots iff

$$2kD(w - D) > N(1 + w - D)^2 \tag{8.35}$$

Otherwise they are real. A limit in which the condition (8.35) is *not* fulfilled is $N \gg 1$. More precisely, in the limit

$$N \to \infty, \qquad k \to \infty, \qquad N/k = \mathcal{O}(1), \qquad w - 1 = \mathcal{O}(1) \tag{8.36}$$

the steady state average population D and the modal intensity $\mathcal{I}$ are given by

$$D = 1 + \frac{w - 1}{2wN} + \mathcal{O}(1/N^2), \qquad \mathcal{I} = \frac{w - 1}{N} + \mathcal{O}(1/N^2) \tag{8.37}$$

Therefore the condition (8.35) becomes

$$N < N_{\text{sup}} \equiv 2k(w - 1)/w^2 \tag{8.38}$$

The interesting feature of this result is that it indicates the existence of a cut-off mode number N_{sup} such that if $N > N_{\text{sup}}$ the roots of $P(2, \lambda) = 0$ are both real and there is no frequency Ω_L. In this regime only the relaxation oscillation Ω_R remains because the cubic is

$$\lambda^3 + \lambda^2 w + \lambda k(w - 1) + kw(w - 1) = 0 \tag{8.39}$$

and its roots are $\lambda_1 = -w$ and $\lambda_{R\pm} = \pm i\Omega_R$. This result is made possible by the fact that in the limit (8.36) the dominant order coefficients of the cubic are independent of the mode number.

8.4.2 Large pump rate limit: One frequency

Another limit in which the condition (8.35) is not satisfied is

$$k \to \infty, \qquad w \to \infty, \qquad k/w = \mathcal{O}(1), \qquad N = \mathcal{O}(1) \qquad (8.40)$$

Then the condition (8.35) becomes

$$w < w_c = 4k/(2N - 1) \qquad (8.41)$$

and the roots of $P(2, \lambda) = 0$ are real if $w > w_c$. In that limit, the cubic (8.34) is

$$\lambda^3 + \lambda^2 w + \lambda kw + kw^2 = 0 \qquad (8.42)$$

which is the large w limit of (8.39). Its three roots are $\lambda_1 = -w$ and $\lambda_{2,3} = \pm i(kw)^{1/2}$. The frequency $(kw)^{1/2}$ is the large w limit of Ω_R.

8.4.3 Linewidth inequalities

If $k \gg 1$ but $N = \mathcal{O}(1)$ and $w - 1 = \mathcal{O}(1)$, it is simple to expand the roots of the cubic (8.34) in powers of $k^{1/2}$. This yields

$$\lambda_1 = -\Gamma_1 + \mathcal{O}(k^{-1/2}) \qquad (8.43)$$

$$\lambda_{2,3} = \pm i\Omega_R k^{1/2} - \Gamma_R + \mathcal{O}(k^{-1/2}) \qquad (8.44)$$

The exact solution of the quadratic (8.33) is

$$\lambda_{4,5} = \pm i\overline{\Omega}_L k^{1/2} - \Gamma_L \qquad (8.45)$$

where the low frequency is given by

$$\overline{\Omega}_L = \Omega_L \sqrt{1 - \frac{(1 + w - D)^2}{4k\Omega_L^2}} \qquad (8.46)$$

The three decay rates are

$$\Gamma_1 = \frac{[(2N - 1)D - 2N]^2 + 2N}{[4N - (2N - 1)D][2N - (2N - 1)D]}$$

$$\Gamma_R = \frac{(2N + 1)[(D + 2N)^2 - 2N(D^2 + 2N + 1)]}{[4N - (2N - 1)D][4N - (4N - 2)D]}$$

$$\Gamma_L = \frac{D}{4N - (4N - 2)D} \qquad (8.47)$$

This gives the linewidth of the peaks in the power spectrum at $\Omega = 0$, Ω_R and $\overline{\Omega}_L$, respectively. From these expressions, a simple relation between the three decay rates is obtained.

$$\Gamma_R = 2\Gamma_L - \frac{1}{2}\Gamma_1 \qquad (8.48)$$

It may be difficult to measure experimentally Γ_1 because $\Omega = 0$ is not necessarily in the same range of detection as the two nonzero frequencies. For that case, an inequality may be derived between Γ_R and Γ_L

$$\Gamma_R < 2\Gamma_L \qquad (8.49)$$

8.4.4 Close to threshold: Zero, one, or two frequencies

Another limit that bears interesting results is the vicinity of the lasing threshold. Let us consider the limit

$$k \rightarrow \infty, \qquad N = \mathcal{O}(1), \qquad w = 1 + \eta, \qquad \eta = \mathcal{O}(1/k) \quad (8.50)$$

Using the results (8.16), it is easy to derive an expansion for the five roots of the characteristic equation (8.34). The cubic becomes

$$\lambda^3 + 2\lambda^2 + \lambda(1 + k\eta) + k\eta = 0 \qquad (8.51)$$

As in Section 8.4.1, the dominant order coefficients of the cubic are independent of the number of modes. The solutions of (8.47) are

$$\lambda_1 = -1, \qquad \lambda_{R\pm} = \frac{1}{2}\left(-1 \pm \sqrt{1 - 4k\eta}\right) \qquad (8.52)$$

It follows from these expressions that there is a critical excess pump η defined by

$$\eta_R = \frac{1}{4k} \qquad (8.53)$$

If $\eta < \eta_R$, the three roots (8.52) are real and negative. If $\eta > \eta_R$, the roots $\lambda_{R\pm}$ are complex conjugate and their imaginary part is the relaxation oscillation frequency that is given by $\Omega_R^2 = \eta k - 1/4 = k(w - 1) - 1/4$. Away from the lasing threshold we recover the usual expression $\Omega_R^2 = k(w - 1)$.

In the limit (8.50), the degenerate quadratic $P(2, \lambda) = 0$ has the solutions

$$\lambda_{L\pm} = \frac{1}{2}\left(-1 - \frac{2N\eta}{2N + 1} \pm \sqrt{1 - \frac{4k\eta}{2N + 1}}\right) \qquad (8.54)$$

Here again a threshold value appears for the excess pump η. Defining

$$\eta_{L,N} = (2N + 1)/4k \qquad (8.55)$$

we find that the two roots (8.54) are real and negative if $\eta < \eta_{L,N}$. For η larger than $\eta_{L,N}$, the two roots are complex conjugate and the imaginary part gives the oscillation frequency Ω_L.

From this analysis, we can draw the following conclusions on the dynamics of the N mode laser with $k_p = k \gg 1$, $\gamma_p = 1$ and finite N.

1. At threshold, $w = 1$, there are N vanishing roots and $N + 1$ real negative roots.

2. In the domain $1 < w < 1 + 1/4k$, there is no internal oscillation frequency: All roots are real and negative.

3. In the domain $1 + 1/4k < w < 1 + (2N + 1)/4k$, there is only the Ω_R oscillation frequency.

4. In the domain $1 + (2N + 1)/4k < w < 4k/(2N - 1)$, the two oscillation frequencies Ω_R and Ω_L exist.

5. In the domain $w > 4k/(2N - 1)$, only the Ω_R relaxation frequency remains.

8.4.5 No degeneracy: N frequencies

To close this chapter, let us recall that the two models on which we have based our analysis are degenerate. The assumptions of equal gains for the N modes lead to a unique low-frequency relaxation oscillation Ω_L. The assumption of equal losses leads to identical line widths Γ_L. If $\gamma_p \neq 1$ but $k_p = k$, the polynomial $[P(2, \lambda)]^{N-1}$ becomes $P(2N - 2, \lambda)$, which may generate $N - 1$ different low-frequency relaxation oscillations Ω_{Lp}. In that case, the conclusion of Section 8.4.4 remains qualitatively correct provided we replace the unique threshold $w = 1 + \eta_{L,N}$ by the N thresholds $w_p = 1/\gamma_p + \eta_{L,p}$. Thus, each time the threshold of oscillation of a mode is reached, there are two new roots, one of which vanishes at the threshold. Slightly above threshold, the new roots become complex conjugate and determine the emergence of a new relaxation oscillation [12].

References

[1]   K. Otsuka, P. Mandel, S. Bielawski, D. Derozier, and P. Glorieux, *Phys. Rev. A* **46** (1992) 1692.

[2]   K. Otsuka, M. Georgiou, and P. Mandel, *Jpn. J. Appl. Phys.* **31** (1992) L1250.

[3]   K. Otsuka, P. Mandel, M. Georgiou, and C. Etrich, *Jpn. J. Appl. Phys.* **32** (1993) L318.

[4]   N. B. Abraham, L. L. Everett, C. Iwata, and M. B. Janicki, *SPIE Proc.* **2095** (1994) 16.

[5]   K. Wiesenfeld, C. Bracikowski, G. James, and R. Roy, *Phys. Rev. Lett.* **65** (1990) 1749.

[6]   E. A. Viktorov, D. R. Klemer, and M. A. Karim, *Opt. Commun.* **113** (1995) 441.

[7]   C. Bracikowski and R. Roy, *Chaos* **1** (1991) 49; *Phys. Rev. A* **43** (1991) 6455.

[8]   J.-Y. Wang and P. Mandel, *Phys. Rev. A* **48** (1993) 671.

[9]   P. Mandel and J.-Y. Wang, *Optics Lett.* **19** (1994) 533.

[10]   J.-Y. Wang, P. Mandel, and T. Erneux, *Quantum & Semiclass. Opt.* **7** (1995) 461.

[11]   K. Otsuka, P. Mandel, and J.-Y. Wang, *Opt. Commun.* **112** (1994) 71.

[12]   P. Mandel, M. Georgiou, K. Otsuka, and D. Pieroux, *Opt. Commun.* **100** (1993) 341.

[13]   D. Pieroux and P. Mandel, *Opt. Commun.* **107** (1994) 245.

[14]   B. A. Nguyen and P. Mandel, *Opt. Commun.* **112** (1994) 235.

9

Laser stability

Laser theory has attracted a large number of studies centered on the stability problem [1]–[4]. There are at least three motivations for these studies. First, the laser equations are rather simple equations that can be derived from first principles with a minimal admixture of phenomenology. The complexity of their solutions was not fully appreciated until Haken showed the equivalence between the three single-mode laser equations on resonance [equations (1.58)–(1.60) with $\Delta = 0$] and the Lorenz equations derived in hydrodynamics [5]. However, the domains of parameters that are relevant for optics and hydrodynamics are not the same. New asymptotic studies were suggested for the laser equations. Second, infrared lasers have been built that can be modeled quite accurately by the two-level equations, at least in some domains of parameters. A good analysis of this topic is found in [6]. A marked advantage of optics over hydrodynamics is that in general the time scales are much shorter. Hence experimental data can be accumulated, and averaging procedures can be used to separate the effect of noise from deterministic properties. Third, the laser stability becomes an essential question when the laser is used as a tool in scientific or industrial applications.

Stability means that a perturbation applied to the laser decreases in time. This requires the solution of an initial value problem that in general is too difficult to be solved analytically. In Section 4.4.1 we were able to carry through this analysis on a simple model problem. But this is exceptional. More generally, the analysis proceeds by way of a linearization around the deviation from the state whose stability is probed. The point we want to stress, however, is that stability does not refer to steady solutions only. Periodic and quasi-periodic solutions are also either stable or unstable. The stability of chaotic solutions can also be discussed, but this requires a more subtle definition of the stability concept. In this chapter, we concentrate on the stability of the finite intensity steady state of the laser equations. In Chapter 10 the stability of periodic solutions will be analyzed.

9.1 Unidirectional single-mode ring laser

9.1.1 Formulation of the stability equation

Our starting point is the unidirectional single-mode ring laser equations (1.48)–(1.50) derived in the first chapter. We can easily remove the assumption that all atoms are identical and contribute in the same way to the macroscopic polarization. If there is a distribution $f(\omega_a)$ of atomic frequencies, the total polarization that was introduced in Chapter 1 as the single-atom polarization, $p = \mu(\mathcal{P} + \mathcal{P}^*)$, multiplied by the number of atoms, N, becomes the single atom polarization $p(\omega)$ integrated over all frequencies with the distribution $f(\omega)$. All the steps leading from (1.12)–(1.14) to (1.48)–(1.50) remain unaffected by this generalization. To simplify the analysis, we assume that the nonlinear medium fills the resonant cavity so that $L = \ell$ and that the laser is operating in the single-mode regime. Using the physical time t, the Maxwell–Bloch equations (1.48)–(1.50) in a fixed reference frame ($\overline{\omega} = \omega_c$) are

$$\partial E(t)/\partial t = -(\gamma_c + i\omega_c)E(t) - \gamma_c A \int_0^{+\infty} P(\omega, t)f(\omega)\, d\omega \qquad (9.1)$$

$$\partial P(\omega, t)/\partial t = -(\gamma_\perp + i\omega)P(\omega, t) - \gamma_\perp E(t)D(\omega, t) \qquad (9.2)$$

$$\partial D(\omega, t)/\partial t = \gamma_\parallel\{1 - D(\omega, t) + (1/2)[E^*(t)P(\omega, t) + E(t)P^*(\omega, t)]\} \qquad (9.3)$$

Let us introduce the decomposition

$$E(t) = \mathcal{E}(\tau)\exp[-i\varphi(\tau)], \qquad P(\omega, t) = [\mathcal{P}(\omega, \tau) + i\mathcal{Q}(\omega, \tau)]\exp[-i\varphi(\tau)] \qquad (9.4)$$

where the functions $\mathcal{E}$, $\mathcal{Q}$, $\mathcal{P}$, and φ are real and time is scaled with respect to the cavity decay rate: $\tau = \gamma_c t$. These functions satisfy the equations

$$\partial\mathcal{E}(\tau)/\partial\tau = -\mathcal{E}(\tau) - A\int_0^{+\infty} \mathcal{P}(\omega, \tau)f(\omega)\, d\omega \qquad (9.5)$$

$$\mathcal{E}(\tau)\delta = A\int_0^{+\infty} \mathcal{Q}(\omega, \tau)f(\omega)\, d\omega \qquad (9.6)$$

$$\kappa\, \partial\mathcal{Q}(\omega, \tau)/\partial\tau = -\mathcal{Q}(\omega, \tau) + \Delta(\omega)\mathcal{P}(\omega, \tau) \qquad (9.7)$$

$$\kappa\, \partial\mathcal{P}(\omega, \tau)/\partial\tau = -\mathcal{P}(\omega, \tau) - \Delta(\omega)\mathcal{Q}(\omega, \tau) - D(\omega, \tau)\mathcal{E}(\tau) \qquad (9.8)$$

$$(\kappa/\gamma)\, \partial D(\omega, \tau)/\partial\tau = -D(\omega, \tau) + 1 + \mathcal{E}(\tau)\mathcal{P}(\omega, \tau) \qquad (9.9)$$

where two new detunings have been introduced

$$\delta = \partial\varphi/\partial\tau - \omega_c/\gamma_c, \qquad \Delta(\omega) = \kappa\, \partial\varphi/\partial\tau - \omega/\gamma_\perp \qquad (9.10)$$

with $\kappa = \gamma_c/\gamma_\perp$ and $\gamma = \gamma_\parallel/\gamma_\perp$. From these equations it appears that there are four differential equations for the field and atomic variables amplitudes $\mathcal{E}$, $\mathcal{P}$, $\mathcal{Q}$, and D and one algebraic equation for the time derivative of the field phase. This last relation is the dispersion relation, which gives, in steady state, the laser frequency.

To progress one step further, we restrict the analysis to atomic frequency distributions that are symmetric around a maximum and assume that the cavity is tuned to that maximum. Thus $f(\omega) \equiv f(|\omega_c - \omega|)$. Typical distributions are the gaussian and the lorentzian functions. Because ω_c is in the optical domain, $f(0)$ is vanishingly small and $f(-|\omega_c - \omega|) < f(0)$. For this reason, we extend the lower limit of the integration over the frequencies to $-\infty$. With these assumptions and approximations, it is easy to verify that there exists a class of solutions with $\partial\varphi/\partial\tau = \omega_c/\gamma_c$

$$\partial\mathcal{E}(\tau)/\partial\tau = -\mathcal{E}(\tau) - A \int_{-\infty}^{+\infty} \mathcal{P}(\omega, \tau) f(\omega)\, d\omega \tag{9.11}$$

$$\kappa\, \partial\mathcal{Q}(\omega, \tau)/\partial\tau = -\mathcal{Q}(\omega, \tau) + \Delta(\omega)\mathcal{P}(\omega, \tau) \tag{9.12}$$

$$\kappa\, \partial\mathcal{P}(\omega, \tau)/\partial\tau = -\mathcal{P}(\omega, \tau) - \Delta(\omega)\mathcal{Q}(\omega, \tau) - D(\omega, \tau)\mathcal{E}(\tau) \tag{9.13}$$

$$(\kappa/\gamma)\, \partial D(\omega, \tau)/\partial\tau = -D(\omega, \tau) + 1 + \mathcal{E}(\tau)\mathcal{P}(\omega, \tau) \tag{9.14}$$

with $\Delta(\omega) = (\omega_c - \omega)/\gamma_\perp$. The steady state is defined by the condition that the four amplitudes are time-independent, whereas the phase of the field is of the form $\varphi(\tau) = \varphi(0) + \Omega_\ell \tau$. Solving the equations for the three material variables in steady state yields

$$\mathcal{P}(\omega) = -\mathcal{E}/[1 + \Delta^2(\omega) + I] \tag{9.15}$$

$$\mathcal{Q}(\omega) = -\mathcal{E}\Delta(\omega)/[1 + \Delta^2(\omega) + I] \tag{9.16}$$

$$D(\omega) = [1 + \Delta^2(\omega)]/[1 + \Delta^2(\omega) + I] \tag{9.17}$$

where the intensity of the oscillating mode is $I = \mathcal{E}^2$. The properties of the zero intensity solution have already been analyzed in the first chapter. Hence we consider in this chapter solutions with $I \neq 0$. With this restriction, the steady state equations for the field amplitude and frequency are given by

$$1 = A \int_{-\infty}^{+\infty} \frac{f(\omega)}{1 + \Delta^2(\omega) + I}\, d\omega \tag{9.18}$$

$$\delta = -A \int_{-\infty}^{+\infty} \frac{\Delta(\omega)f(\omega)}{1 + \Delta^2(\omega) + I}\, d\omega \tag{9.19}$$

To determine the stability of this steady state, we seek solutions of the evolution equations (9.11)–(9.14) of the form $[\mathcal{E}(\tau), \mathcal{Q}(\omega, \tau), \mathcal{P}(\omega, \tau), D(\omega, \tau)] = [\mathcal{E}, \mathcal{Q}(\omega), \mathcal{P}(\omega), D(\omega)] + \varepsilon[e(\tau), q(\omega, \tau), p(\omega, \tau), d(\omega, \tau)] + \mathcal{O}(\varepsilon^2)$. To first order in ε, we obtain a set of linear differential equations with constant coefficients. Hence their solution is of the form $[e(\tau), q(\omega, \tau), p(\omega, \tau), d(\omega, \tau)] = [e, q(\omega), p(\omega), d(\omega)] \exp(\lambda\tau)$. The system of differential equations being homogeneous, the algebraic system of equations for the coefficients e, $q(\omega)$, $p(\omega)$, and $d(\omega)$ has nontrivial solutions only if the determinant of the coefficients vanishes. This leads to the characteristic equation

$$1 + \lambda = A(1 + \lambda\kappa) \int_{-\infty}^{+\infty} f(\omega)R(\omega)[(1 + \lambda\kappa/\gamma)D(\omega) + \mathcal{E}P(\omega)]\, d\omega$$

$$(9.20)$$

$$1/R(\omega) = \Delta^2(\omega)(1 + \lambda\kappa/\gamma) + (1 + \lambda\kappa)[(1 + \lambda\kappa/\gamma)(1 + \lambda\kappa) + I]$$

$$\equiv (1 + \lambda\kappa/\gamma)[\Gamma + \Delta^2(\omega)] \qquad (9.21)$$

$$\Gamma = (1 + \lambda\kappa)^2 + I(1 + \lambda\kappa)/(1 + \lambda\kappa/\gamma) \qquad (9.22)$$

Using the expression (9.15) for $P(\omega)$, we finally obtain the characteristic equation in the form

$$(1 + \lambda)\frac{1 + \lambda\kappa/\gamma}{1 + \lambda\kappa} = A \int_{-\infty}^{+\infty} f(x)\frac{(1 + \lambda\kappa/\gamma)(1 + x^2) - I}{(\Gamma + x^2)(1 + I + x^2)}\, dx \quad (9.23)$$

where $x = \Delta(\omega)$. Since the stability is directly governed by the roots of this equation, we sometimes refer to (9.23) as the stability equation.

9.1.2 Lorentzian broadening on resonance

At this point in the development of the stability theory, we need to specify the distribution function $f(x)$. As stated earlier, the two standard choices are the lorentzian and the gaussian distributions. In this chapter we choose the lorentzian distribution

$$f(x) = \frac{1}{\pi}\frac{u}{u^2 + x^2}, \qquad \int_{-\infty}^{+\infty} f(x)\, dx = 1 \qquad (9.24)$$

The width of the distribution $f(\omega)$ is called the *inhomogeneous width*, as opposed to the homogeneous (or natural) width $\gamma_\perp$ associated with the spontaneous decay. With the scaling adopted in this chapter, the function $f(x)$ has a width u that is the inhomogeneous width divided by the homogeneous width. Two limits are especially relevant.

1. The homogeneous limit defined by $u \to 0$. In this limit, $f(x) \to \delta(x)$ and we recover the results of Chapter 1.

2. The inhomogeneous limit defined by $u \to +\infty$.

The important property is that the integral in the stability equation (9.23) has the same expression in the inhomogeneous limit whether we use the lorentzian distribution (9.24) or a gaussian distribution with a width proportional to u. This was verified by two independent evaluations of the stability equation in the inhomogeneous limit and published in the same journal, same issue [7, 8]. Since $u = 0$ and $u = \infty$ are the only two limits in which analytic results have been obtained so far, it does not really matter which of the two distributions we use. Therefore our choice is the lorentzian distribution because it still gives reasonably simple analytical expressions for an arbitrary width u. This is not the case with the gaussian distribution. Furthermore, the use of the gaussian distribution raises the question of the center-of-mass distribution. Most gas lasers operating in the visible are pumped by an electric discharge, either directly or indirectly. It is not clear that in such conditions an equilibrium velocity distribution is a valid choice.

The steady state equation for the laser intensity is given by (9.18)

$$1 = \frac{A}{(u + \sqrt{1 + I})\sqrt{1 + I}} \tag{9.25}$$

The physical solution is

$$I = -1 + \frac{1}{4}\left(-u + \sqrt{u^2 + 4A}\right)^2 \tag{9.26}$$

In the homogeneous limit, we recover $I = A - 1$ and the lasing threshold is $A_{\text{th}} = 1$. In the inhomogeneous limit the dominant contribution to the intensity is $I = -1 + (A/u)^2$ and the lasing threshold is $A_{\text{th}} = u$.

Using the implicit equation (9.25) for the intensity, a straightforward calculation leads from the general stability equation (9.23) to the implicit equation

$$\frac{1 + \lambda\kappa/\gamma}{1 + \lambda\kappa}\lambda(1 - \kappa)$$

$$= [(1 + \lambda\kappa/\gamma)(1 - \Gamma) - I]\frac{u + \sqrt{\Gamma} + \sqrt{1 + I}}{(u + \sqrt{\Gamma})(\sqrt{\Gamma} + \sqrt{1 + I})\sqrt{\Gamma}} \tag{9.27}$$

The remarkable fact is that this implicit equation for λ can be reduced without any approximation to a polynomial of finite degree despite the integral nature of the general equation (9.23). We consider successively the homogeneous and the inhomogeneous limits.

In the *homogeneous limit* ($u = 0$), the characteristic equation reduces to the cubic

$$\lambda^3 \kappa^2 + \lambda^2 \kappa(\kappa + \gamma + 1) + \lambda\gamma(\kappa + A) + 2\gamma(A - 1) = 0 \qquad (9.28)$$

The correspondence with the results derived in Section 1.5 is easy to establish. Because of the different scaling of the time ($\tau = \kappa T$), the λ of Section 1.5 becomes $\lambda\kappa$ in this chapter. On resonance, the quintic (1.71) factors into $\lambda\kappa(\lambda\kappa + \kappa + 1)$ times the cubic (9.28). This cubic has a real root that vanishes at $A = 1$, the laser first threshold. The other possibility of instability is a Hopf bifurcation in which the real part of a pair of complex conjugate roots changes sign. At this laser second threshold, the finite intensity steady state becomes unstable and a periodic solution emerges from that point. To find this bifurcation point, we seek solutions of the cubic polynomial (9.28) in the form $(\lambda^2 + \Omega_H^2)(\lambda + \rho) = 0$. This leads to

$$\rho = 1 + (1 + \gamma)/\kappa, \qquad \Omega_H^2 = \frac{2\gamma(\kappa + 1)}{\kappa(\kappa - 1 - \gamma)}, \qquad A_H = \frac{\kappa(\kappa + \gamma + 3)}{\kappa - 1 - \gamma}$$

$$(9.29)$$

and A_H is the pump parameter at which the bifurcation occurs. Thus, the second threshold exists only in the so-called bad cavity limit $\kappa > 1 + \gamma$. In terms of unscaled decay rates, the bad cavity limit is $\gamma_c > \gamma_\perp + \gamma_\parallel$. The oscillation frequency Ω_H is unrelated to the optical lasing frequency Ω_ℓ. The frequency Ω_H is an oscillation frequency that is damped if $A < A_H$ and amplified if $A > A_H$. The Hopf bifurcation is the limit in which the damping of a relaxation oscillation vanishes. On resonance, this Hopf bifurcation leads always to unstable periodic solutions. For $A > A_H$, there is a transition to a chaotic state. The reader is referred to [6] for a good analysis of the behavior of the laser and of the laser equations around the Hopf bifurcation. Additional numerical results are be found in [9] and [10].

In the limit where both atomic decay rates are equal, $\gamma = 1$, more analytic results can be derived. The frequency and the intensity at the Hopf bifurcation are

$$\Omega_H^2 = 2(\kappa + 1)/[\kappa(\kappa - 1)], \qquad I_H = (\kappa + 1)(\kappa + 2)/(\kappa - 2) \qquad (9.30)$$

For $\kappa \gg 2$, the critical intensity I_H diverges linearly with κ. For $0 < \kappa - 2 \ll 1$, the critical intensity diverges like $1/(\kappa - 2)$. The minimum of the critical intensity is $7 + 2\sqrt{12} \simeq 13.93$ that is reached for $\kappa = 2 + \sqrt{12} \simeq 5.46$.

In order to keep the algebra within reasonable limits, we analyze the *inhomogeneous limit* ($u \rightarrow \infty$) with equal atomic decay rates: $\gamma = 1$. With this simplification, the stability equation (9.27) becomes the quintic

$$\sum_{n=0}^{5} a(n)\lambda^n = 0 \tag{9.31}$$

$a(5) = \kappa^3$

$a(4) = 2\kappa^2(\kappa + 1)$

$a(3) = -\kappa^3 I + \kappa^2(4 + 2I) + \kappa(1 + I)$

$a(2) = 2\kappa(1 + 3I)$

$a(1) = \kappa I(2 - I) + 2I(1 + I)$

$a(0) = 2I^2$

The derivation of this quintic requires a little sleight of hand. From the stability equation (9.27) one first obtains an equation of the form $\Gamma^{1/2}F = (1 + I)^{1/2}G$, where F and G are simple polynomials in λ. Squaring this equation to obtain a polynomial in λ, we risk introducing spurious roots that are solutions of $\Gamma^{1/2}F = -(1 + I)^{1/2}G$ and not of the original equation. It turns out that there is one such root, $\lambda = -2/\kappa$. The quintic (9.31) is obtained after elimination of this spurious root.

The quintic has the expected steady bifurcation at $I = 0$ corresponding to the laser first threshold. To check for the presence of a second threshold, we seek solutions of this quintic in the form $\lambda^2 + \Omega_H^2$ times a cubic. This provides an expression of the frequency at the Hopf bifurcation

$$\Omega_H^2 = \frac{I_H}{\kappa} \frac{I_H(\kappa^2 - 2) - 2(\kappa + 1)^2}{I_H(\kappa^3 - \kappa^2 - 1) - (2\kappa + 1)^2} \tag{9.32}$$

where the intensity I_H is the real root of the cubic

$$I_H^3 b(3) + I_H^2 b(2) + I_H b(1) + b(0) = 0 \tag{9.33}$$

$b(3) = (\kappa - 1)^5$

$b(2) = -7\kappa^4 + 14\kappa^3 - 4\kappa^2 - 4$

$b(1) = 6\kappa^4 + 12\kappa^3 - 10\kappa^2 - 15\kappa - 5$

$b(0) = -2(2\kappa + 1)^2(\kappa + 1) \tag{9.34}$

Therefore the Hopf bifurcation exists for $\kappa > 1$ if $u \to \infty$ instead of the $\kappa > 2$ condition if $u = 0$. In the bad cavity limit ($\kappa \gg 1$), the roots of the quintic at the Hopf bifurcation are

$\lambda_{H1} = -2 + \mathcal{O}(1/\kappa), \qquad \lambda_{H2,3} = \pm i/\kappa + \mathcal{O}(1/\kappa^2)$

$\lambda_{H4} = -4/3\kappa + \mathcal{O}(1/\kappa^2), \qquad \lambda_{H5} = -4/3\kappa^2 + \mathcal{O}(1/\kappa^3) \tag{9.35}$

Hence the oscillation frequency Ω_H is $1/\kappa$ and the intensity is $I_H = 4/3\kappa + \mathcal{O}(1/\kappa^2)$. The main conclusion of this analysis is that the critical intensity at which the laser becomes unstable is significantly reduced by the inhomogeneous broadening. This conclusion was first obtained by Casperson [11, 12] in his seminal work on laser stability.

9.1.3 Nonresonant steady state intensity

All the results obtained until now rely on two assumptions: the resonance condition $\partial\varphi/\partial\tau = \omega_c/\gamma_c$ and the symmetry of the frequency distribution function $f(\omega) \equiv f(|\omega_c - \omega|)$. It is not possible to fully assess to what extent these restrictions affect the stability results. Out of resonance, the phase of the laser field is coupled to the amplitude; phase instabilities have been reported. However, there has been no report of the phase instability happening for lower pump parameters than the amplitude instability. Although much energy has been devoted to this topic, it does not seem to lead to any conclusive result. However, one example where the detuning has a drastic influence is on the steady state intensity. To understand this point, let us again consider the steady state equations and the lorentzian function but without the resonance assumption.

$$1 = A \int_{-\infty}^{+\infty} \frac{f(\omega)}{1 + \Delta^2(\omega) + I} \, d\omega \tag{9.36}$$

$$\delta = -A \int_{-\infty}^{+\infty} \frac{\Delta(\omega)f(\omega)}{1 + \Delta^2(\omega) + I} \, d\omega \tag{9.37}$$

$$f(\omega) = \frac{1}{\pi} \frac{u}{u^2 + (\tilde{\omega} - \omega)^2/\gamma_\perp^2} \tag{9.38}$$

The lorentzian distribution is peaked around $\omega = \tilde{\omega} \neq \omega_c$. The steady state frequency and intensity are given by

$$1 = \frac{A}{\sqrt{1+I}} \frac{u + \sqrt{1+I}}{\Delta^2 + [u + \sqrt{1+I}]^2}, \qquad \delta = \frac{-A\Delta}{\Delta^2 + [u + \sqrt{1+I}]^2} \tag{9.39}$$

In these expressions, $\Delta \equiv \Delta(\tilde{\omega}) = (\Omega_\ell - \tilde{\omega})/\gamma_\perp$ and $\delta = (\Omega_\ell - \omega_c)/\gamma_c$, where Ω_ℓ is the lasing frequency. At the laser first threshold, $I = 0$ by definition. Hence the first threshold is characterized by

$$A_1 = (u + 1)\left[1 + \left(\frac{\delta_{\text{ca}}}{1 + u + \kappa}\right)^2\right] \tag{9.40}$$

$$\Delta_1 = -\delta_{ca}\frac{1+u}{1+u+\kappa}, \qquad \delta_1 = \delta_{ca}/(1+u+\kappa) \qquad (9.41)$$

where $\delta_{ca} = (\omega_c - \tilde{\omega})/\gamma_\perp$ is the atom-field detuning.

With these results we construct the finite intensity solution in the vicinity of the first threshold. We define a vicinity of the threshold by $A = A_1 + \varepsilon$ with $|\varepsilon| \ll 1$, and we seek a solution of the steady state equations (9.39) in power series of ε : $I = \varepsilon J + \mathcal{O}(\varepsilon^2)$, $\Delta = \Delta_1 + \mathcal{O}(\varepsilon)$, and $\delta = \delta_1 + \mathcal{O}(\varepsilon)$. This leads to

$$J = \frac{2}{u}\frac{(1+u+\kappa)^3}{\kappa-u-1}\left[\left(1+\frac{2}{u}\right)\frac{(1+u+\kappa)^3}{\kappa-u-1} - \delta_{ca}^2\right] \qquad (9.42)$$

Because $I > 0$, J and ε must have the same sign. If the slope is negative, we have a case of bistability because for A sufficiently large we recover from (9.39) $\partial I/\partial A > 0$. The condition for bistability is therefore $J < 0$, which leads to

$$\kappa > 1 + u, \qquad \delta_{ca}^2 > \delta_{crit}^2 \equiv \left(1+\frac{2}{u}\right)\frac{(1+u+\kappa)^3}{\kappa-u-1} \qquad (9.43)$$

The critical value δ_{crit} has a minimum at $\kappa = 2(u+1)$, where its value is

$$\min(\delta_{crit}^2) = 27(u+1)^2(1+2/u) \qquad (9.44)$$

Finally, let us simply mention that from equations (9.39) a quartic can be obtained for the auxiliary variable $K = \sqrt{1+I}$. Indeed, combining the two equations in (9.39) gives $\delta = -\Delta K/(u+K)$, which can be solved for the lasing frequency Ω_ℓ to give

$$\Delta(\tilde{\omega}) = -\delta_{ca}(u+K)/[u+K(1+\kappa)] \qquad (9.45)$$

Inserting this expression into the first of the two equations (9.39) yields a quartic for K. From this quartic, it can be shown that another domain of bistability occurs at higher intensity [7]. Similar results have also been reported for gaussian broadening [13] though with much less analytical results.

9.2 Unidirectional multimode ring laser

9.2.1 *Formulation of the stability equation*

In the preceding section, we have analyzed the stability of the single-mode steady solution (9.15)–(9.19) against perturbations that preserve the single-mode character of the laser. One may question the wisdom of such a restriction. In principle, nothing prevents fluctuations at other frequencies to be amplified and to destabilize the single-mode solutions (9.15)–(9.19). To test this conjecture (which turns out to be true), we consider the more general equations (1.42)–(1.45) in which the single-mode restriction has not yet been introduced. We proceed along the same line as in Section 9.1.1 to obtain the multimode generalization of these equations

$$\mathcal{E}_T(\tau, \xi) + \mathcal{E}_\xi(\tau, \xi) = -\mathcal{E}(\tau, \xi) - A \int_0^{+\infty} \mathcal{P}(\omega, \tau, \xi) f(\omega) \, d\omega \qquad (9.46)$$

$$\mathcal{E}(\tau, \xi)[\varphi_T(\tau, \xi) + \varphi_\xi(\tau, \xi)] = A \int_0^{+\infty} \mathcal{Q}(\omega, \tau, \xi) f(\omega) \, d\omega \qquad (9.47)$$

$$\kappa \mathcal{Q}_T(\omega, \tau, \xi) = -\mathcal{Q}(\omega, \tau, \xi) + \Delta(\omega) \mathcal{P}(\omega, \tau, \xi) \qquad (9.48)$$

$$\kappa \mathcal{P}_T(\omega, \tau, \xi) = -\mathcal{P}(\omega, \tau, \xi) - \Delta(\omega) \mathcal{Q}(\omega, \tau, \xi) - D(\omega, \tau, \xi) \mathcal{E}(\tau, \xi) \qquad (9.49)$$

$$(\kappa/\gamma) D_T(\omega, \tau, \xi) = -D(\omega, \tau, \xi) + 1 + \mathcal{E}(\tau, \xi) \mathcal{P}(\omega, \tau, \xi) \qquad (9.50)$$

using the notation $g_T \equiv \partial g / \partial \tau$ and $g_\xi \equiv \partial g / \partial \xi$ for any function g. The scaling of the space–time variables is $\tau = \gamma_c t$ and $\xi = \gamma_c z / c$ with $\ell = L$. To these equations we must add the ring cavity boundary conditions

$$Z(\xi + \kappa L/v, \tau) = Z(\xi, \tau)$$

$$Z(\xi, \tau) = \{\mathcal{E}(\xi, \tau), \mathcal{Q}(\omega, \xi, \tau), \mathcal{P}(\omega, \xi, \tau), D(\omega, \xi, \tau)\}$$

$$\varphi(\xi + \kappa L/v, \tau) = \varphi(\xi, \tau) + 2\pi n, \qquad n = 0, \pm 1, \pm 2, \ldots \qquad (9.51)$$

For the frequency distribution, we use the lorentzian distribution (9.38). The steady state solution for the material amplitudes is still given by (9.15)–(9.17). The field amplitude and phase in steady state are given by

$$\mathcal{E}_\xi = -\mathcal{E} + \frac{A\mathcal{E}}{\sqrt{1+I}} \frac{u + \sqrt{1+I}}{\Delta^2 + [u + \sqrt{1+I}]^2}$$

$$\varphi_T + \varphi_\xi = -\frac{A\Delta}{\Delta^2 + [u + \sqrt{1+I}]^2}, \qquad \varphi_{TT} = 0 \qquad (9.52)$$

where $\Delta = \kappa \varphi_T - \tilde{\omega}/\gamma_\perp$ and $I = \mathcal{E}^2$. A steady state solution that is compatible with these equations is

$$\mathcal{E}_\xi = 0, \qquad \varphi_T = -\varphi_\xi = \tilde{\omega}/\gamma_\perp, \qquad I = -1 + \frac{1}{4}\left(-u + \sqrt{u^2 + 4A}\right)^2$$

$$(9.53)$$

This is the resonant solution whose stability against single-mode perturbations was analyzed in Section 9.1.2. To assess the stability of this solution against multimode perturbations, we seek solutions of equations (9.46)–(9.50) in the form $Z(\xi, \tau) = Z(\xi) + \varepsilon\delta Z(\xi, \tau) + \mathcal{O}(\varepsilon^2)$ for the amplitudes $\mathcal{E}$, $\mathcal{Q}$, $\mathcal{P}$, and D, with $Z(\xi)$ being the steady state solution derived from (9.53). For the phase of the field, the corresponding expansion is $\varphi(\xi, \tau) = (\tau - \xi)\tilde{\omega}/\gamma_\perp + \varepsilon\delta\varphi(\xi, \tau) + \mathcal{O}(\varepsilon^2)$. Introducing these expansions into the evolution equations (9.46)–(9.50), we linearize the resulting equations with respect to ε. This leads to a set of coupled linear partial differential equations with constant coefficients. Hence the solutions of these equations are $\delta Z(\xi, \tau) = \delta Z \exp(i\alpha\xi + \lambda\tau)$ and $\delta\varphi(\xi, \tau) = \delta\varphi \exp(i\alpha\xi + \lambda\tau)$. In order to simplify the algebra, we take again the limit $\gamma = 1$. The four amplitude equations yield the characteristic equation

$$\lambda + i\alpha + 1 = A(\lambda\kappa + 1)(J_0 + J_2) - IAJ_2 \tag{9.54}$$

The phase yields the characteristic equation

$$1 + i\alpha/\lambda + \kappa A\Gamma J_0/(1 + \lambda\kappa) - \kappa AJ_2 = 0 \tag{9.55}$$

where the functions J_0 and J_2 are defined as

$$J_n = \int_{-\infty}^{+\infty} \frac{\Delta^n f(\omega)}{(1 + \Delta^2 + I)(\Gamma + \Delta^2)} d\omega, \qquad \Gamma = I + (1 + \lambda\kappa)^2 \tag{9.56}$$

so that

$$J_0 = \frac{u + \sqrt{\Gamma} + \sqrt{1 + I}}{\sqrt{\Gamma}(\sqrt{\Gamma} + \sqrt{1 + I})(u + \sqrt{\Gamma})}, \qquad J_2 = \frac{u\sqrt{1 + I}}{(\sqrt{\Gamma} + \sqrt{1 + I})(u + \sqrt{\Gamma})} \tag{9.57}$$

Equations (9.54) and (9.55) are the general stability equations. It is not difficult to derive stability equations at the same level of generality as (9.18) and (9.19). However, such general equations have not been studied for $u \neq 0$.

9.2.2 The Risken–Nummedal instability

In this section we consider the homogeneous limit of the stability equations. Because $J_0 \to 1/\Gamma$ and $J_2 \to 0$ for $u \to 0$, it is simple to verify that the characteristic equation for the phases becomes

$$\lambda^2\kappa + \lambda(1 + \kappa + i\alpha\kappa) + i\alpha = 0 \tag{9.58}$$

whereas the characteristic equation for the amplitudes is

$$\lambda^3\kappa^2 + \lambda^2(\kappa^2 + i\alpha\kappa^2 + 2\kappa) + \lambda(1 + \kappa + I + 2i\alpha\kappa) + 2I + i\alpha(I + 1) = 0 \tag{9.59}$$

These two equations were derived in the same year by Risken and Nummedal [14, 15] and by Graham and Haken [16]. They are referred to in the scientific literature as the *Risken–Nummedal equations* (and the corresponding instability point as the *Risken–Nummedal instability*) because these authors made at the same time a systematic numerical study of the solutions that emerge from the instability point.

For $\alpha = 0$, we recover the characteristic polynomials derived and discussed in Section 9.1.2. Hence we concentrate our attention on the solutions for non-vanishing α. Instabilities with $\alpha \neq 0$ correspond to the emergence of solutions with an explicit space dependence in the propagation direction. Numerical results indicate that these solutions are often pulses that propagate in the laser ring cavity. Since $\lambda = 0$ is not a solution if $\alpha \neq 0$, we seek instability points corresponding to Hopf bifurcations: $\lambda = i\Omega_H$. The phase equation (9.58) does not admit such an instability. However, the amplitude equation (9.59) may lead to a Hopf bifurcation. The real part of the amplitude equation yields the relation

$$\alpha_H = \left[2I_H - \Omega_H^2\kappa(\kappa + 2)\right]/2\kappa\Omega_H \qquad (9.60)$$

Inserting this relation into the imaginary part of the amplitude equation gives a biquadratic equation for the critical frequency Ω_H, whose solutions are

$$\Omega_{H\pm}^2 = \left(3I_H - 1 \pm \sqrt{I_H^2 - 14I_H + 1}\right)/2\kappa^2 \qquad (9.61)$$

Because $\tau = \gamma_c t$, where t is the physical time, the oscillation frequency $\Omega_{H\pm}$ scales like $\gamma_\perp$. If we had not introduced the simplification $\gamma = 1$, we would have obtained the scaling $\Omega_{H\pm} \sim \sqrt{\gamma_\perp\gamma_\parallel}$ on the physical time scale t. This is completely different from the scaling of the relaxation oscillation frequency $\Omega_R \sim \sqrt{\gamma_c\gamma_\parallel}$ obtained in (7.54). The different scaling indicates a difference of physical mechanism. The scaling of Ω_R suggests the combined influence of the cavity losses and the atomic population damping. On the contrary, the scaling of $\Omega_{H\pm}$ depends on the cavity properties only via the steady intensity I. Another difference is that Ω_R can be made arbitrarily small because it also scales like $w - 1$, that is, the normalized difference between the pump parameter and its threshold value.

A critical wave number $\alpha_{H\pm}$ is associated with each critical frequency $\Omega_{H\pm}$. The instability threshold therefore corresponds to the condition $\alpha_{H+} = \alpha_{H-} \equiv \alpha_H$, implying $\Omega_{H+} = \Omega_{H-} \equiv \Omega_H$ and therefore

$$I_H = 7 + 2\sqrt{12} \simeq 13.93 \qquad (9.62)$$

This gives a frequency of the order of $\sqrt{\gamma_\perp\gamma_\parallel I}$ that is comparable to the Rabi frequency. Thus, in the homogeneous limit $u = 0$, the Risken–Nummedal is associated with the excitation of an atomic resonance rather than a laser resonance. Note that $I_H = \mathcal{O}(1)$ so that $\kappa\Omega_H = \mathcal{O}(1)$. Therefore, in the bad cavity limit, the critical wave numbers are $\mathcal{O}(1)$

$$\alpha_H \to (2I_H - \kappa^2\Omega_H^2)/2\kappa\Omega_H + \mathcal{O}(1/\kappa) \qquad \text{as } \kappa \to \infty \qquad (9.63)$$

whereas in the good cavity limit they diverge like $1/\kappa$

$$\alpha_H \to -(\kappa\Omega_H)/\kappa + \mathcal{O}(1) \qquad \text{as } \kappa \to 0 \qquad (9.64)$$

Since $\alpha\zeta = (\alpha\gamma_c/c)z$, the wave numbers α have to be multiplied by γ_c/c to become dimensional physical quantities.

The above analysis rests on the assumption $\gamma = 1$. Did we miss anything with this restriction? Yes, and in fact more than meets the eye. In the homogeneous limit $u = 0$, it is quite easy to derive the stability equation for an arbitrary value of γ. The critical frequency (9.61) then becomes

$$\Omega_{H\pm}^2 = \gamma\left(3I - \gamma \pm \sqrt{I^2 - 2I(4 + 3\gamma) + \gamma^2}\right)/2\kappa^2 \qquad (9.65)$$

The only reason we have introduced the restriction $\gamma = 1$ is that this is the only case for which analytic results have been obtained in the inhomogeneous limit and we wish to keep a unified presentation. However, we have already discussed in Section 7.1 the problem of the decay rate inequalities, and we have explained that for a large class of lasers the natural limit is $\gamma_\parallel \ll \gamma_c \ll \gamma_\perp$ or $\gamma \ll \kappa \ll 1$. This corresponds to the rate equation limit. Because in this limit $I_H = 8 + \mathcal{O}(\gamma)$ and therefore $\Omega_H^2 = \mathcal{O}(\gamma/\kappa^2)$, we still have to specify the magnitude of the ratio γ/κ^2 before any statement can be made on the critical frequency and the associated wave number. Clearly, Ω_H can diverge, remain finite, or tend to zero without violating the condition that γ and κ are small parameters. These different cases lead to quite different solutions. An asymptotic analysis of the Risken–Nummedal instability in the rate equation limit has been recently published [17, 18]; it shows how to deal correctly with these problems.

To close this section, we should emphasize the difference between the single mode second threshold instability (9.29), which exists only in a bad cavity, and the Risken–Nummedal instability, which can occur in either a good or a bad cavity. Despite this last property, it still has to be identified experimentally.

9.2.3 The inhomogeneous limit

In the inhomogeneous limit, the functions J_0 and J_2 have simple expressions

$$J_0 \to 1/\sqrt{\Gamma}(\sqrt{\Gamma} + \sqrt{1 + I}), \qquad J_2 \to \sqrt{1 + I}/(\sqrt{\Gamma} + \sqrt{1 + I}) \qquad (9.66)$$

Let us first consider the amplitude stability that is ruled by the equation (9.54). In order to obtain an algebraic polynomial in λ, we have to square an equation to remove the square roots. As in the single-mode case, this procedure introduces the spurious root $\lambda = -2/\kappa$. Once this root has been removed, we are left with the quintic

$$\sum_{n=0}^{5} c(n)\lambda^n = 0 \qquad (9.67)$$

$c(5) = \kappa^3$

$c(4) = 2\kappa^2(1 + \kappa + i\alpha\kappa)$

$c(3) = -\kappa^3 I + \kappa^2(4 + 2I) + \kappa(1 + I) - \alpha^2\kappa^3 + 2i\alpha\kappa^2(2 + \kappa)$

$c(2) = 2\kappa(1 + 3I) - 2\alpha^2\kappa^2 + 2i\alpha\kappa(2\kappa + 1) + 2i\alpha\kappa(1 + \kappa)I$

$c(1) = \kappa I(2 - I) + (2I - \alpha^2\kappa)(1 + I) + 2i\alpha\kappa(1 + 3I)$

$c(0) = 2I^2 + 2i\alpha I(1 + I)$

Apart from the laser first threshold at $I = 0$, the other possible mechanism of instability is a Hopf bifurcation for which $\lambda = i\Omega_H$. Inserting this expression into the imaginary part of (9.67) leads to a relation between the wave number α_H, the frequency Ω_H, and intensity I_H at the critical point

$$\alpha_H = -\Omega_H + \frac{I_H}{\kappa\Omega_H} \frac{\kappa^4\Omega_H^4 + 2\kappa^2\Omega_H^2(1 - I_H) + I_H(1 + I_H)}{\kappa^4\Omega_H^4 + 2\kappa^2\Omega_H^2(1 - I_H) + (1 + I_H)^2} \qquad (9.68)$$

When this relation is introduced into the real part of (9.67), we obtain an equation for the frequency at the Hopf bifurcation in the form of another quintic

$$\sum_{n=0}^{5} d(n)(\kappa^2\Omega_H^2)^n = 0 \qquad (9.69)$$

$d(5) = 1$

$d(4) = 3 - 5I_H$

$d(3) = 1 - 5I_H + 10I_H^2$

$d(2) = -3 + 4I_H - 3I_H^2 - 10I_H^3$

$d(1) = -2 - 5I_H + I_H^2 + 9I_H^3 + 5I_H^4$

$d(0) = I_H^2(1 + I_H)^2(2 + I_H)$

For $I_H = 0$, the five roots of (9.69) are $0, 1, -1, -1$, and -2. Hence already for $I_H = 0$ we find a solution, $\Omega_H\kappa = 1$, that is indeed a solution of the original equation and not a spurious root. Thus in the inhomogeneous limit, the Hopf bifurcation coincides with the laser first threshold and is characterized by the frequency and wave number

$$\Omega_H = 1/\kappa, \qquad \alpha_H = -1/\kappa, \qquad I_H = 0 \qquad (9.70)$$

On the physical space and time scales, the frequency equals γ_T and the wave number becomes $-\gamma_\perp/c$. This is the same scaling as for the Risken–Nummedal instability. The difference is the intensity dependence that has disappeared in the inhomogeneous limit. It should be borne in mind that the result (9.70) is not an exact result but gives only the dominant term in a series expansion in powers of $1/u$.

9.3 Bidirectional single-mode ring laser

9.3.1 Formulation and steady states

At this point in our stability analysis of a ring laser, it is time to consider a problem that is long overdue: Why do we always consider unidirectional propagation? In this section we remove this assumption. Following the type of decomposition used in Chapter 7, we consider solutions of equations (1.18)–(1.20) that are of the form

$$E(z, t) = \{E_1(t)\exp[i(kz - \omega t)] + E_2(t)\exp[i(kz + \omega t)] + c.c.\}/2 \qquad (9.71)$$

$$P(z, t) = i\{P_1(z, t)\exp[i(kz - \omega t)] + P_2(z, t)\exp[i(kz + \omega t)]\}/2 \qquad (9.72)$$

where ω and k are the cavity frequencies and wave number ω_c and k_c introduced in the first chapter. The field equation (1.18) yields

$$\partial E_1/\partial t = E_2 e^{2i\omega t} - \frac{N\mu\omega}{2\varepsilon_0 L} \int_0^L \left(P_1 + P_2 e^{2i\omega t} + P_1^* e^{-2i(kz-\omega t)} + P_2^* e^{-2ikz} \right) dz \qquad (9.73)$$

Because ω and k are the optical frequency and wave number, we expect that the amplitudes E_j and P_j are slowly varying functions of time and that the P_j are also slowly varying functions of space. On this basis we neglect all terms in (9.73) that are affected by a time-periodic exponential. We also neglect for the same reason the polarization amplitude that is affected by a space-periodic exponential *in the integral over space*. Adding to the resulting equation the phenomenological linear damping leads to

$$\partial E_j(t)/\partial t = -\gamma_c E_j(t) - \frac{N\mu\omega}{2\varepsilon_0 L} \int_0^L P_j(z, t)\, dz \qquad (9.74)$$

with $j = 1$ or 2. For obvious reasons, z and t are still the physical unscaled space and time variables. An element we have also introduced is that the cavity decay rate γ_c is the same for the two modes. Thus the cavity is perfectly symmetric with respect to both directions of propagation.

For the two material equations, we neglect terms affected by an oscillation in time and obtain

$$\partial P_j(z, t)/\partial t = (i\omega - i\omega_a - \gamma_\perp)P_j(z, t) - \frac{\mu}{\hbar}D(z, t)[E_j(t) + E_{3-j}^*(t)e^{-2ikz}] \tag{9.75}$$

$$\partial D(z, t)/\partial t = -\gamma[D(z, t) - D_a] + \frac{\mu}{2\hbar}[E_1(t)P_1^*(z, t) + E_2(t)P_2^*(z, t) + c.c.] \tag{9.76}$$

We restrict our analysis to the resonant case $\omega = \omega_a$ for which a more adequate decomposition of the fields and polarizations is in terms of the quadratures

$$E_j = X_j + iY_j, \qquad P_j = U_j + iV_j \tag{9.77}$$

To remain consistent with the previous sections of this chapter, we also introduce the scaled time $\tau = \gamma_c t$. The dynamical equations for the new variables are

$$\partial X_j(\tau)/\partial \tau = -X_j(\tau) - \frac{N\omega\mu}{2\gamma_c\varepsilon_0 L}\int_0^L U_j(z, \tau)\,dz \tag{9.78}$$

$$\partial Y_j(\tau)/\partial \tau = -Y_j(\tau) - \frac{N\omega\mu}{2\gamma_c\varepsilon_0 L}\int_0^L V_j(z, \tau)\,dz \tag{9.79}$$

$$\kappa\,\partial U_j(z, \tau)/\partial \tau = -U_j(z, \tau) - \frac{\mu}{\hbar\gamma_\perp}D(z, \tau)[X_j + X_{3-j}\cos(2kz)$$
$$- Y_{3-j}\sin(2kz)] \tag{9.80}$$

$$\kappa\,\partial V_j(z, \tau)/\partial \tau = -V_j(z, \tau) - \frac{\mu}{\hbar\gamma_\perp}D(z, T)[Y_j + Y_{3-j}\cos(2kz)$$
$$- X_{3-j}\sin(2kz)] \tag{9.81}$$

$$(\kappa/\gamma)\,\partial D(z, \tau)/\partial \tau = D_a - D(z, \tau) + \frac{\mu}{\hbar\gamma_\parallel}\sum_j[X_j U_j + Y_j V_j] \tag{9.82}$$

with $\kappa = \gamma_c/\gamma_\perp$ and $\gamma = \gamma_\parallel/\gamma_\perp$.

The analysis of the steady state solutions is made simpler if we realize that the phase of the fields is arbitrary in steady state. Therefore, we can set $Y_j = 0$ without loss of generality and we obtain the steady state solutions of the material equations (9.80)–(9.82) in the form

$$D(z) = D_a/\left\{1 + \frac{\mu^2}{\hbar^2\gamma_\parallel\gamma_\perp}[X_1^2 + X_2^2 + 2X_1 X_2\cos(2kz)]\right\} \tag{9.83}$$

$$U_j(z) = -\frac{\mu}{\hbar\gamma_\perp} D(z)[X_j + X_{3-j}\cos(2kz)] \tag{9.84}$$

$$V_j(z) = \frac{\mu}{\hbar\gamma_\perp} D(z)X_{3-j}\sin(2kz) \tag{9.85}$$

The population grating is manifest in the expression (9.83) for the population inversion. The structure of this expression also indicates that the grating results from the interference between the two counterpropagating modes. If either mode has a vanishing intensity, the grating vanishes and we are back to the usual single-mode unidirectional ring laser theory. Using the property that $kL/2\pi$ is an integer, the steady state equation for the two modal intensities $I_j = X_j^2 \mu^2/(\hbar^2\gamma_\perp\gamma_\|)$ is

$$2I_j = A\{1 - (1 - I_j + I_{3-j})/B\} \tag{9.86}$$

where $B^2 \equiv 1 + 2(I_1 + I_2) + (I_1 - I_2)^2$ and the pump parameter is defined as usual through $A = N\mu^2\omega D_a/(2\hbar\gamma_c\gamma_\perp\varepsilon_0)$. There are three types of steady state solutions:

1. The trivial solution $I_1 = I_2 = 0$
2. The two unidirectional solutions $I_j = A - 1, I_{3-j} = 0$ for $j = 1$ or 2
3. The bidirectional solution $I_1 = I_2 = \{4A - 1 - \sqrt{8A + 1}\}/8$

The three nonzero solutions emerge from the trivial solution at the same bifurcation point $A = 1$. Therefore a stability analysis is required to determine which solution, if any, is stable above the laser first threshold.

9.3.2 Linear stability

We proceed in two steps to clarify the analysis. In the first part of this section, no assumption will be made on the decay rates γ_c, $\gamma_\perp$, and $\gamma_\|$ while performing the linear stability analysis of the steady states. It turns out that this lack of specification does not affect the stability status of the solutions. In the second part of this section, we analyze the nature of the roots of the characteristic equations. For this purpose, explicit orders of magnitude will be assigned to the decay rates.

The linear stability of the trivial solution $I_1 = I_2 = 0$ yields the characteristic equation

$$(\lambda\kappa + \gamma)[(\lambda\kappa + 1)(\lambda + 1) - A]^4 = 0 \tag{9.87}$$

This ninth-degree polynomial has only negative roots for $A < 1$ and a fourfold degenerate positive root for $A > 1$. Hence the trivial solution is stable below the laser first threshold $A = 1$ and unstable above it.

The linear stability of the symmetric solution $I_1 = I_2 \neq 0$ also gives a ninth-degree polynomial that factors into a quadratic and a quintic. The quadratic in turn factors into

$$\lambda(\lambda\kappa + \kappa + 1)[(\lambda\kappa + 1)(\lambda + 1) - 1 - 4I_1] = 0 \qquad (9.88)$$

The root $\lambda = 0$ is associated with the neutral stability of the absolute phase of the fields. The steady state intensities do not depend on the phases that determine the Y_j. Likewise, the linear stability analysis *on resonance* is not affected by the specific choice of the phases. This invariance implies a neutral stability with respect to a change of the phase, a property that is translated by the existence of the zero root. The main point, however, is that the quadratic equation in (9.88) has always a positive root, which implies that the symmetric solution is always unstable in the physical domain $I \geq 0$.

Finally, we consider the stability of the unidirectional solution. The two asymmetric solutions have the same stability domain. The initial condition alone determines which of the two solutions is excited. Here again the stability is ruled by a polynomial of degree nine that factors into the quadratic equation $\lambda(\lambda\kappa + \kappa + 1) = 0$, a cubic

$$\lambda^3\kappa^2 + \lambda^2\kappa(\kappa + \gamma + 1) + \lambda\gamma(\kappa + A) + 2\gamma(A - 1) = 0 \qquad (9.89)$$

and a quartic

$$\lambda^4\kappa^3 + \lambda^3\kappa^2(\kappa + \gamma + 2) + \lambda^2\kappa[A\gamma + (\gamma + 1)(\kappa + 1)]$$
$$+ \lambda(3\kappa A + 2A - \kappa)\gamma/2 + \gamma(A - 1) = 0 \qquad (9.90)$$

The cubic is identical to (9.28), which rules the stability of the oscillating mode. The quartic rules the stability of the suppressed mode. If $A = 1 + \varepsilon$ with $\varepsilon > 0$, the two roots $-\gamma/\kappa$ and $-(1 + \kappa)/\kappa$ are common to both equations. In addition, the cubic has the root $-2\varepsilon/(\kappa + 1)$, whereas the quartic has the roots $-\kappa$ and $-\varepsilon/(2\kappa + 2)$. Thus, the asymmetric solutions are stable near the laser first threshold. Sufficiently far from the first threshold, a second threshold occurs in the form of a Hopf bifurcation. Each of the two polynomial equations (9.89) and (9.90) provides a Hopf bifurcation if the bad cavity condition $\gamma_c > \gamma_\parallel + \gamma_\perp$ (or $\kappa > \gamma + 1$) is satisfied. Which of the two Hopf bifurcations has the lowest pump parameter (and therefore occurs first) depends on the relative magnitudes of the three decay rates. An analysis of this point has not been made yet.

An interesting property of the roots that determine the stability of the asymmetric solution appears in the rate equation limit (7.10)

$$\gamma \ll \kappa \ll 1, \qquad A = \mathcal{O}(1) \qquad (9.91)$$

It is clear that A is also assumed to be smaller than the critical value at which a Hopf bifurcation takes place. In this limit, the dominant contribution to the three roots associated with the finite intensity mode are

$$\lambda_1 = -1/\kappa, \qquad \lambda_{2,3} = \pm i \sqrt{2(\gamma/\kappa)(A - 1)} \qquad (9.92)$$

and the dominant contribution to the four roots associated with the zero intensity mode are

$$\lambda_{1,2} = -1/\kappa, \qquad \lambda_{3,4} = \pm i \sqrt{(\gamma/\kappa)(A - 1)} \qquad (9.93)$$

Because the time is scaled as $\tau = \gamma_c t$, we have $\lambda_{2,3}\tau = \pm i \sqrt{2\gamma_{\parallel}\gamma_c(A - 1)}t$ and the relaxation oscillation introduced at the end of Chapter 7 is recovered for the oscillating mode. The only difference between the damped oscillation frequencies of the two modes is the factor $\sqrt{2}$, a property that has been verified experimentally.

References

[1]   N. B. Abraham, L. A. Lugiato and L. M. Narducci, eds., *Instabilities in active optical media, J. Opt. Soc. Am. B* **2** (1985) 1–264.

[2]   R. W. Boyd, M. G. Raymer, and L. M. Narducci, eds., *Optical Instabilities*, Cambridge studies in modern optics **4**, 262–264 (Cambridge University Press, Cambridge, 1986).

[3]   D. K. Bandy, A. N. Oraevsky, and J. R. Tredicce, eds., *Nonlinear dynamics of lasers, J. Opt. Soc. Am. B* **5** (1988) 876–1215.

[4]   N. B. Abraham, P. Mandel, and L. M. Narducci, *Dynamical Instabilities and Pulsations in Lasers*, Progress in optics Vol. XXV, pp 1–190, E. Wolf ed. (North-Holland, Amsterdam, 1988).

[5]   H. Haken, *Phys. Lett.* **53A** (1975) 77.

[6]   C. O. Weiss and R. Villaseca, *Dynamics of Lasers* (Physik-Verlag, Weinheim, 1991).

[7]   P. Mandel, *J. Opt. Soc. Am. B* **2** (1985) 112.

[8]   J. Y. Zhang, H. Haken, and H. Ohno, *J. Opt. Soc. Am. B* **2** (1985) 141.

[9]   H. Zeghlache and P. Mandel, *J. Opt. Soc. Am. B* **2** (1985) 18.

[10] P. Mandel, A. C. Maan, B. J. Verhaar, and H. T. C. Stoof, *Phys. Rev. A* **44** (1991) 608.

[11] L. W. Casperson, *Phys. Rev. A* **21** (1980) 911.

[12] L. W. Casperson, *Phys. Rev. A* **23** (1981) 248.

[13] N. B. Abraham, L. A. Lugiato, P. Mandel, L. M. Narducci, and D. K. Bandy, *J. Opt. Soc. Am. B* **2** (1985) 35.

[14] H. Risken and K. Nummedal, *J. Appl. Phys.* **39** (1968) 4662.

[15] H. Risken and K. Nummedal, *Phys. Lett. A* **26** (1968) 275.

[16] R. Graham and H. Haken, *Z. Physik* **213** (1968) 420.

[17] T. W. Carr and T. Erneux, *Phys. Rev. A* **50** (1994) 724.

[18] T. W. Carr and T. Erneux, *Phys. Rev. A* **50** (1994) 4219.

10

Second harmonic generation

10.1 Introduction

Up to now, we have described in detail many properties of steady bifurcations and limit points, that is, critical points where a stable steady state solution loses its stability and coincides with another steady state solution. At a few places, we have also met the so-called Hopf bifurcation where a steady solution loses its stability and a time-periodic solution emerges. However, we have not yet studied in any detail a Hopf bifurcation for lack of a suitable example. Even the simple-looking trio of laser equations on resonance [equations (1.58)–(1.60) with $\Delta = 0$, E and P real] yield such complex expressions that it is hard to separate conceptual difficulties from mere computational problems. In this chapter, we make an intrusion upon a domain that has not yet been considered in this book. The motivation is both to cover an important topic of cavity nonlinear optics and to provide a pedagogical example of a Hopf bifurcation.

In the preceding chapters, this book has dealt exclusively with processes in which only one photon is either absorbed or emitted. Other phenomena, however, rely on multiphoton transitions [1, 2]. In the original Bohr formulation of atomic transitions and in much of the ensuing quantum mechanical formulation, resonance conditions on atomic transitions express conservation laws but give no constraint on the number of photons needed to achieve the transition. The simplest multiphoton processes are two-photon transitions. We will analyze two frequency conversion processes that do not conserve photon number. The first process is second harmonic generation (SHG), in which two photons with frequency ω are absorbed by a nonlinear medium and only one photon, with frequency 2ω, is emitted. The converse process, optical parametric amplification, is the absorption of one photon of frequency ω and the emission of two photons whose frequencies add up to ω. Momentum conservation leads to phase matching conditions. The other two-photon processes that preserve the photon number are scattering processes of which the Raman scattering is the best known example. We mostly consider SHG in this chapter. Some aspects of parametric amplification are analyzed in Chapter 11.

10.2 Formulation

The conversion of two photons of the same frequency into one photon with twice that frequency is a highly nonlinear process. It requires a medium that operates the conversion. The difficulty is that from the Maxwell–Bloch equations (1.48)–(1.50) it follows that the material polarization of a two-level medium is of the form $P \sim E(a_0 + a_2|E|^2 + a_4|E|^4 + \cdots)$. Thus the two-level model used so far is inadequate to explain SHG because this requires a contribution to the polarization that is quadratic in the electric field. However, if the system that must mediate the SHG lacks the reflexion symmetry (such as noncentrosymmetric crystals), the diagonal matrix elements ν_j of the electric dipole in equation (1.3) do not necessarily vanish. The result is a modification of the Bloch equation (1.19) for the complex atomic polarization amplitude that becomes

$$\partial \mathcal{P}/\partial t = -[\gamma_\perp + i\omega_a + i(\nu_1 - \nu_0)E/\hbar]\mathcal{P} - (i\mu/\hbar)E\mathcal{D} \qquad (10.1)$$

This suggests the following field dependence of the atomic polarization

$$P = E[a_0 + a_1E + b_1E^* + \mathcal{O}(E^2)] \qquad (10.2)$$

To use that piece of information, let us consider Maxwell's equation (1.1)

$$c^2\,\partial^2\mathbf{E}/\partial z^2 - \partial^2\mathbf{E}/\partial t^2 = \varepsilon_0^{-1}\,\partial^2\mathbf{P}/\partial t^2 \qquad (10.3)$$

We have kept the assumption of one-dimensional propagation though we have to introduce the vector nature of the electric field and of the polarization

$$\mathbf{E} = \sum_{j=1}^{3} \mathbf{e}_j E_j, \qquad \mathbf{P} = \sum_{j=1}^{3} \mathbf{e}_j P_j \qquad (10.4)$$

where the $\{\mathbf{e}_j\}$ are pairwise orthogonal unit vectors. Because we wish to study cavity SHG, we expand the field and polarization components onto a complete set of modes. We use the unidirectional running modes so that, strictly speaking, the modal expansion and the remainder of this chapter apply only to ring lasers. However, whenever the grating induced by the field spatial variation is negligible, the equations derived in this chapter are also fair approximations to describe Fabry–Pérot lasers. Finally, we note that the use of unidirectional running modes also covers SHG in free space without cavity. Hence we introduce the decomposition

$$E_j = \frac{1}{2}\sum_\alpha \left(E_{j\alpha}F(j\alpha) + c.c.\right), \qquad P_j = \frac{1}{2}\sum_\alpha \left(P_{j\alpha}F(j\alpha) + c.c.\right)$$

$$F(j\alpha) = \exp[i(k_{j\alpha}z - \omega_{j\alpha}t)] \qquad (10.5)$$

The function $F(j\alpha)$ takes away the fast optical space–time variation of the field and polarization modal amplitudes. However, due to the light–matter interaction, there remains a residual space and time dependence in both $E_{j\alpha}$ and $P_{j\alpha}$. The convention used in this chapter is that Roman indices refer to the three cartesian space coordinates and the Greek indices refer to the mode expansion.

At this point we have to supplement Maxwell's equation with a relation between the P_j's and the E_j's. We assume the following vectorial and multimode generalization of (10.2)

$$P_{j\alpha}F(j\alpha) = \epsilon_0\chi^{(1)}_{j\alpha}E_{j\alpha}F(j\alpha) + \sum_{pq}\sum_{\beta\gamma}\chi^{(2a)}_{jpq\alpha\beta\gamma}E_{p\beta}E_{q\gamma}F(p\beta)F(q\gamma)$$

$$+ \sum_{pq}\sum_{\beta\gamma}\chi^{(2b)}_{jpq\alpha\beta\gamma}E_{p\beta}E^*_{q\gamma}F(p\beta)F^*(q\gamma)$$

$$+ \sum_{pqr}\sum_{\alpha\beta\gamma}\chi^{(3a)}_{jpqr\alpha\beta\gamma\delta}E_{p\beta}E_{q\gamma}E_{r\delta} + \cdots \qquad (10.6)$$

The $\chi^{(1)}$, $\chi^{(2)}$, and $\chi^{(3)}$ are related to the linear, second-order and third-order susceptibilities, respectively. They are arbitrary as of now and it would require a model of the medium with which the light interacts to derive a microscopic expression of these coefficients. The Bloch equations (1.19) and (1.20) are such a model, though inadequate for SHG in crystals. Thus, based on experimental evidence, we assume that the coefficients in expansion (10.6) can be different from zero and we analyze the consequences of that assumption.

The expansion (10.6) is far from being general. In fact, it is clear from the discussions of adiabatic elimination all over this book that the expansion (10.6) requires the material variables to relax much faster than any field variable. Otherwise $\mathbf{P}(t)$ would have to depend on $\mathbf{E}(t')$ integrated over the domain $0 \le t' \le t$. The linear susceptibility $\chi^{(1)}_{j\alpha}$ leads to the linear dispersion relation

$$c^2k_{j\alpha}^2 = \omega_{j\alpha}^2(1 + \chi^{(1)}_{j\alpha}), \qquad (10.7)$$

the linear refractive index $n_{j\alpha}$, the linear dielectric function $\varepsilon_{j\alpha}$ and the speed of light in the medium $v_{j\alpha}$

$$n_{j\alpha} = \sqrt{1 + \chi^{(1)}_{j\alpha}}, \qquad \varepsilon_{j\alpha} = \varepsilon_0(1 + \chi^{(1)}_{j\alpha}), \qquad v_{j\alpha} = c/n_{j\alpha} \qquad (10.8)$$

The index α in these equations expresses the frequency dependence of n, ε and v. The experimental situation that we want to analyze in this chapter is the following. A nonlinear crystal, able to generate SHG, is placed into a cavity that can be made resonant for the fundamental and the second harmonic frequencies (doubly resonant cavity). A monochromatic source is used to generate a field oscillating at the fundamental frequency. This field is injected into the cavity where the up-conversion is realized.

We now proceed to derive the evolution equations for the components $E_{j\alpha}$ of the electric field inside the cavity. We insert the expansion (10.6) and retain only the linear and quadratic contributions in the complex field amplitude. This leads to an equation of the form

$$\sum_j \sum_\alpha \mathbf{e}_j \left(\overline{\mathcal{E}}_{j\alpha}(z, t) F(\alpha j) + \overline{\mathcal{E}}^*_{j\alpha}(z, t) F^*(j\alpha) \right) = 0 \qquad (10.9)$$

where

$$\overline{\mathcal{E}}_{j\alpha}(z, t) \equiv \partial E_{j\alpha}/\partial t \pm v_{j\alpha} \partial E_{j\alpha}/\partial z$$
$$- \frac{i}{2n^2_{j\alpha}\epsilon_0 \omega_{j\alpha}} \sum_{pq} \sum_{\beta\gamma} \left(\chi^{(2a)}_{jpq\alpha\beta\gamma}(\omega_{p\beta} + \omega_{q\gamma})^2 E_{p\beta} E_{q\gamma} F(p\beta) F(q\gamma) \right.$$
$$\left. + \chi^{(2b)}_{jpq\alpha\beta\gamma}(\omega_{p\beta} - \omega_{q\gamma})^2 E_{p\beta} E^*_{q\gamma} F(p\beta) F^*(q\gamma) \right) F^*(j\alpha) \qquad (10.10)$$

To derive this equation, we have used the dispersion relation (10.7) and we have made the slowly varying envelope approximation by assuming that inequalities (6.2) for the field hold. Obviously, all powers higher than the second have been neglected in $\overline{\mathcal{E}}_{j\alpha}$. If we only retain in $\overline{\mathcal{E}}_{j\alpha}$ resonant processes, for which either of the triple products of modes $F(p\beta)F(q\gamma)F^*(j\alpha)$ and $F(p\beta)F(q\gamma)F^*(j\alpha)$ equal unity, we can use the same argument as in (7.16) to cancel each function $\overline{\mathcal{E}}_{j\alpha}$ separately. This yields

$$\partial E_{j\alpha}/\partial t \pm v_{j\alpha} \partial E_{j\alpha}/\partial z =$$
$$\frac{i}{2n^2_{j\alpha}\epsilon_0 \omega_{j\alpha}} \sum_{pq} \sum_{\beta\gamma} \left(\overline{\chi}^{(2a)}_{jpq\alpha\beta\gamma} E_{p\beta} E_{q\gamma} F(p\beta) F(q\gamma) F^*(j\alpha)(\omega_{p\beta} + \omega_{q\gamma})^2 \right.$$
$$\left. + \overline{\chi}^{(2b)}_{jpq\alpha\beta\gamma} E_{p\beta} E^*_{q\gamma} F(p\beta) F^*(q\gamma) F^*(j\alpha)(\omega_{p\beta} - \omega_{q\gamma})^2 \right) \qquad (10.11)$$

The overbar of the susceptibility tensor is a reminder that the equation is derived under the condition that the sums extend over all configurations that make the products of mode functions equal to 1. In the case of SHG, there are only two modes in the cavity. Let the cavity mode 1 oscillate at frequencies $\omega_{j1} = \omega_j$ and the cavity mode 2 oscillate at frequencies $\omega_{j2} = 2\omega_j$. We then have

$$\partial E_{j1}/\partial t \pm v_{j1} \partial E_{j1}/\partial z = \frac{i\omega_{j1}}{2n^2_{j1}\epsilon_0} \sum_{pq} \chi^{(2b)}_{jpq121} E_{p2} E^*_{q1} \theta(p2, -q1, -j1) \qquad (10.12)$$

and

$$\partial E_{j2}/\partial t \pm v_{j2}\,\partial E_{j2}/\partial z = \frac{i\omega_{j2}}{2n_{j2}^2\epsilon_0}\sum_{pq}\chi^{(2a)}_{jpq211}E_{p1}E_{q1}\theta(p1,q1,-j2) \quad (10.13)$$

where we have introduced the functions

$$\theta(p\alpha,\pm q\beta,\pm j\gamma) = \delta(\omega_{p\alpha}\pm\omega_{q\beta}\pm\omega_{j\gamma})\delta(k_{p\alpha}\pm k_{q\beta}\pm k_{j\gamma}) \quad (10.14)$$

To proceed further, we input the experimental fact that SHG takes place with two orthogonally polarized modes [3]. Hence the two evolution equations become

$$\partial E_{j1}/\partial t \pm v_{j1}\,\partial E_{j1}/\partial z = \frac{i\omega_{j1}}{2n_{j1}^2\epsilon_0}\chi^{(2b)}_{jpj121}E_{p2}E_{j1}^*\theta(p2,-j1,-j1)$$

$$(10.15)$$

$$\partial E_{p2}/\partial t \pm v_{p2}\,\partial E_{p2}/\partial z = \frac{i\omega_{p2}}{2n_{p2}^2\epsilon_0}\chi^{(2a)}_{pjj211}E_{j1}^2\theta(j1,j1,-p2) \quad (10.16)$$

with $p \neq j$. The θ functions, which are identical, imply that $\omega_p = \omega_j$ and $k_{p2} = 2k_{j1}$. This last equality gives the phase matching condition $k^{(2\omega)} = 2k^{(\omega)}$.

Finally, the SHG evolution equations can be written as

$$\partial E_{j1}/\partial t \pm v_{j1}\,\partial E_{j1}/\partial z = iaE_{p2}E_{j1}^* \qquad (10.17)$$

$$\partial E_{p2}/\partial t \pm v_{p2}\,\partial E_{p2}/\partial z = ibE_{j1}^2 \qquad (10.18)$$

with

$$a = \omega_j\chi^{(2b)}_{jpj121}/2n_{j1}^2\epsilon_0, \qquad b = \omega_p\chi^{(2a)}_{pjj211}/n_{p2}^2\epsilon_0 \qquad (10.19)$$

When the crystal is not in a cavity and the light beam propagates in free space, the natural way to analyze these equations is by assuming that the fields are no longer time-dependent and to solve the steady propagation problem [3]. In cavity SHG, the situation is exactly the converse. We assume as in (1.47) that the space dependence has been taken care of by the mode expansion (10.5), but the proper time dependence remains to be found. To transform equations (10.17) and (10.18) into equations fit to describe cavity SHG, we proceed as follows:

1. We seek fields E_{p2} and E_{j1} that are space-independent.
2. In each evolution equation we add a linear cavity damping term $-\gamma_c E$.
3. To specify that we consider cavity SHG, we add in (10.17) a source term E_{ext} representing the fraction of the external field that enters in the cavity.

As a result, we obtain the evolution equations

$$\partial E_{j1}/\partial t = -\gamma_{c1}E_{j1} + iaE_{p2}E_{j1}^* + E_{\text{ext}} \qquad (10.20)$$

$$\partial E_{p2}/\partial t = -\gamma_{c2}E_{p2} + ibE_{j1}^2 \qquad (10.21)$$

The absolute phase is arbitrary and is chosen such that the external field amplitude E_{ext} is real. The generalized situation in which the two modes are driven by external fields has been considered as well [4] and leads to some interesting phase diagrams for the steady solutions.

The final step in the formulation of the cavity SHG is to define dimensionless variables

$$E_{j1} = (\gamma_{c2}/\sqrt{ab})\mathcal{E}_1, \qquad E_{p2} = -(i\gamma_{c2}/a)\mathcal{E}_2, \qquad E_{ext} = (\gamma_{c2}^2/\sqrt{ab})\mathcal{E}$$
$$\tau = \gamma_{c2}t, \qquad \gamma = \gamma_{c1}/\gamma_{c2} \qquad (10.22)$$

in terms of which the evolution equations take the simple form

$$\partial\mathcal{E}_1/\partial\tau = -\gamma\mathcal{E}_1 + \mathcal{E}_1^*\mathcal{E}_2 + \mathcal{E} \qquad (10.23)$$
$$\partial\mathcal{E}_2/\partial\tau = -\mathcal{E}_2 - \mathcal{E}_1^2 \qquad (10.24)$$

This is the standard form in which the intracavity SHG is known in the literature.

10.3 Steady state and stability

In this and the following sections, we focus on the ideal converter limit $\gamma = 0$. In that limit, only the second harmonic field $\mathcal{E}_2$ can escape the cavity. Thus, the input to the cavity is the driving field at frequency ω and the output is the field at frequency 2ω.

The steady state solution is easily found to be

$$\mathcal{E}_1 = \mathcal{E}^{1/3}, \qquad \mathcal{E}_2 = -\mathcal{E}^{2/3} \qquad (10.25)$$

The two fields propagate with orthogonal polarizations and are dephased by π. The physical fields E_1 and E_2 are dephased by $\pi/2$ because $E_1/E_2 = -i\mathcal{E}^{-1/3}\sqrt{a/b}$. To analyze the stability of the steady state solution (10.25) we perform a linear stability analysis, seeking time-dependent solutions of the form

$$\mathcal{E}_1(\tau) = \mathcal{E}^{1/3} + \eta r_1(\tau) + \mathcal{O}(\eta^2), \qquad \mathcal{E}_2(\tau) = -\mathcal{E}^{2/3} + \eta r_2(\tau) + \mathcal{O}(\eta^2) \qquad (10.26)$$

Because the steady solutions are real, we seek real solutions $\mathcal{E}_j(\tau)$. Inserting (10.26) into the evolution equations (10.23)–(10.24) and keeping only terms proportional to η leads to the solution $r_j(t) = r_{j1}\exp(\lambda_1\tau) + r_{j2}\exp(\lambda_2\tau)$ with

$$\lambda_{1,2} = \frac{1}{2}\left(-1 - \mathcal{E}^{2/3} \pm \sqrt{1 + \mathcal{E}^{4/3} - 10\mathcal{E}^{2/3}}\right) \qquad (10.27)$$

These two roots are complex, and they always have a negative real part for a positive real driving field amplitude $\mathcal{E}$. Therefore the steady state is stable against infinitesimal perturbations. This conclusion is restricted by the assumption we have introduced to consider only fluctuations that conserve the reality of the amplitudes. In other words, we have neglected phase fluctuations. To determine their influence, we decompose the complex field amplitudes into quadrature components

$$\mathcal{E}_1 = X + iU, \qquad \mathcal{E}_2 = Y + iV \qquad (10.28)$$

These components satisfy the dynamical equations

$$X' = XY + UV + \mathcal{E} \qquad (10.29)$$

$$Y' = -Y - X^2 + U^2 \qquad (10.30)$$

$$U' = XV - UY \qquad (10.31)$$

$$V' = -V - 2UX \qquad (10.32)$$

A linear stability of the steady solution $X = \mathcal{E}^{1/3}, Y = -\mathcal{E}^{2/3}, U = V = 0$ leads to four characteristic roots. The first pair is again (10.27), and the other pair is

$$\lambda_{3,4} = \frac{1}{2}\left(-1 + \mathcal{E}^{2/3} \pm \sqrt{1 + \mathcal{E}^{4/3} - 6\mathcal{E}^{2/3}}\right)$$

$$= \frac{1}{2}\left(-1 + \mathcal{E}^{2/3} \pm \sqrt{(\mathcal{E}^{2/3} - 3 + 2\sqrt{2})(\mathcal{E}^{2/3} - 3 - 2\sqrt{2})}\right) \qquad (10.33)$$

These two roots are real and negative for $0 \leq \mathcal{E}^{2/3} \leq 3 - 2\sqrt{2} \approx 0.1716$. They are complex conjugate in the domain $3 - 2\sqrt{2} \leq \mathcal{E}^{2/3} \leq 3 + 2\sqrt{2} \approx 5.828$. Above that domain the two roots are again real but with opposite signs. The main point, however, is that the real part of $\lambda_{3,4}$ is negative for $\mathcal{E} < 1$, vanishes at $\mathcal{E} = 1$, and is positive for $\mathcal{E} > 1$. At $\mathcal{E} = 1$, the four roots are

$$\lambda_{1,2H} = -1 \pm i\sqrt{2}, \qquad \lambda_{3,4H} = \pm i \qquad (10.34)$$

Thus at the bifurcation point there is an undamped frequency $\text{Im}(\lambda_H) \equiv \Omega_{3,4H} = 1$ and a damped frequency $\Omega_{1,2H} = \sqrt{2}$ with unit damping rate. Frequencies and damping rates have to be multiplied by γ_{c2} to be expressed in the physical time scale t. In the long time limit, only the undamped oscillations remain and therefore we end up with a periodic solution that emerges from the bifurcation point. Such a bifurcation point, where a steady solution loses its stability for a periodic solution, is known as a *Hopf bifurcation*. By performing the stability analysis first with the real amplitudes and later with the complex amplitudes, we have been able to assign without ambiguity the roots $\lambda_{1,2}$ to the stability of the amplitudes and the roots $\lambda_{3,4}$ to the phase stability. Thus the Hopf bifurcation of cavity SHG is a phase instability. Of course, once the phases become unstable, the phase-sensitive nonlinear coupling $\mathcal{E}_1^* \mathcal{E}_2$ and

$\mathcal{E}_1^2$ induces an instability of the amplitudes. The fact that we deal with a phase instability has important consequences in squeezing, which is maximized in the quadrature that does not cause the bifurcation. Therefore, in SHG the squeezing affects the amplitudes. We now describe how the periodic solution may be constructed and how its stability may be assessed.

10.4 Analysis of the Hopf bifurcation

Since we are dealing with a fairly simple example of Hopf bifurcation, we can explain sequentially how to proceed to determine the solution and its stability. We first integrate the dynamical equations (10.29)–(10.32) in a direct way. This will motivate the multiple time scale analysis of the Hopf bifurcation and finally the solvability condition will be introduced.

10.4.1 Straightforward integration

To perform this direct integration, we first define a small parameter η through

$$\mathcal{E} = 1 + a\eta^2, \qquad a = \pm 1 \tag{10.35}$$

The deviations from the steady state $X = (1+a\eta^2)^{1/3} + x$, $Y = -(1+a\eta^2)^{2/3} + y$, $U = u$ and $V = v$ satisfy the equations

$$x' = -x(1 + a\eta^2)^{2/3} + y(1 + a\eta^2)^{1/3} + xy + uv + a\eta^2 \tag{10.36}$$

$$y' = -y - 2x(1 + a\eta^2)^{1/3} - x^2 + u^2 \tag{10.37}$$

$$u' = v(1 + a\eta^2)^{1/3} + u(1 + a\eta^2)^{2/3} + vx - uy \tag{10.38}$$

$$v' = -v - 2u(1 + a\eta^2)^{1/3} - 2ux \tag{10.39}$$

Equation (10.36) suggests that the dominant effect of the external field $a\eta^2$ will be captured by solutions x and y proportional to η^2. Furthermore, the phase–amplitude coupling will also be captured at the same order if uv is proportional to η^2. These two remarks lead us to seek solutions defined by the expansions

$$x(\tau, \eta) = \eta^2 x_2(\tau) + \mathcal{O}(\eta^4)$$
$$y(\tau, \eta) = \eta^2 y_2(\tau) + \mathcal{O}(\eta^4)$$
$$u(\tau, \eta) = \eta u_1(\tau) + \eta^3 u_3(\tau) + \mathcal{O}(\eta^5)$$
$$v(\tau, \eta) = \eta v_1(\tau) + \eta^3 v_3(\tau) + \mathcal{O}(\eta^5) \tag{10.40}$$

It is important to stress that this expansion holds for all times, including the initial condition. This restriction on the initial condition is often forgotten in numerical simulations of the full equations (10.29)–(10.32). The scaling (10.40) is compatible with equations (10.36)–(10.39) and does not suggest any difficulty.

Inserting these expansions into the evolution equations (10.29)–(10.32) leads to a linear problem at each order in η.

The first-order equations are

$$u_1' = u_1 + v_1, \qquad v_1' = -v_1 - 2u_1 \qquad (10.41)$$

They can be transformed into harmonic oscillator equations $z'' + z = 0$ for u_1 and v_1. Their solution is therefore

$$\begin{pmatrix} u_1 \\ v_1 \end{pmatrix} = b_1 \begin{pmatrix} 1 \\ i - 1 \end{pmatrix} e^{i\tau} - b_1^* \begin{pmatrix} -1 \\ i + 1 \end{pmatrix} e^{-i\tau} \qquad (10.42)$$

The remainder of this subsection is devoted to the determination of the amplitude of oscillations b_1.

The second-order problem is

$$x_2' = y_2 - x_2 + u_1 v_1 + a, \qquad y_2' = -y_2 - 2x_2 + u_1^2 \qquad (10.43)$$

The solution of these equations is

$$\begin{pmatrix} x_2 \\ y_2 \end{pmatrix} = b_2 \begin{pmatrix} 1 \\ i\sqrt{2} \end{pmatrix} e^{(-1+i\sqrt{2})\tau} + b_2^* \begin{pmatrix} 1 \\ -i\sqrt{2} \end{pmatrix} e^{-(1+i\sqrt{2})\tau}$$

$$+ \frac{b_1^2}{1 - 4i} \begin{pmatrix} 2+i \\ -3 \end{pmatrix} e^{2i\tau} + \frac{(b_1^*)^2}{1 + 4i} \begin{pmatrix} 2-i \\ -3 \end{pmatrix} e^{-2i\tau} + \begin{pmatrix} a/3 \\ 2|b_1|^2 - 2a/3 \end{pmatrix} \qquad (10.44)$$

This solution does not give any indication on the amplitude b_1. Therefore we have to consider the next order in the perturbation expansion, which leads to

$$u_3' = u_3 + v_3 + (P_3 e^{3i\tau} + P_1 e^{i\tau} + c.c.) \qquad (10.45)$$

$$v_3' = -v_3 - 2u_3 + (Q_3 e^{3i\tau} + Q_1 e^{i\tau} + c.c.) \qquad (10.46)$$

with

$$P_3 = -b_1^3(24 + 11i)/17, \qquad P_1 = b_1[a(1+i)/3 + |b_1|^2(5i - 20)/17]$$

$$Q_3 = 2b_1^3(2 - 9i)/17, \qquad Q_1 = -b_1[2a/3 + 2|b_1|^2(9i - 2)/17] \quad (10.47)$$

Equations (10.45)–(10.46) are easily solved once they are written as

$$u_3'' + u_3 = e^{3i\tau}[(1 + 3i)P_3 + Q_3] + e^{i\tau}[(1 + i)P_1 + Q_1] + c.c. \quad (10.48)$$

$$v_3'' + v_3 = e^{3i\tau}[-2P_3 + (3i - 1)Q_3] + e^{i\tau}[-2P_1 + (i - 1)Q_1] + c.c. \qquad (10.49)$$

Each equation describes a harmonic oscillator that is forced periodically at the frequencies $\Omega = 1$ and $\Omega = 3$. All the difficulty comes from the fact that the unforced oscillator oscillates at frequency $\Omega = 1$. This resonance causes the solutions of (10.48)–(10.49) to diverge

$$u_3 = u_{31}\tau e^{i\tau} + u_{33}e^{3i\tau} + c.c., \qquad v_3 = v_{31}\tau e^{i\tau} + v_{33}e^{3i\tau} + c.c. \qquad (10.50)$$

$$u_{31} = [(1 + i)P_1 + Q_1]/2i, \qquad u_{33} = -[(1 + 3i)P_3 + Q_3]/8$$

$$v_{31} = (i - 1)u_{31}, \qquad v_{33} = [2P_3 + (1 - 3i)Q_3]/8 \qquad (10.51)$$

10.4.2 Multiple time scale expansion

An explicit evaluation of the troublesome coefficient gives

$$u_{31} = b_1[a(1 + i)/3 + |b_1|^2(21i - 33)/34] \qquad (10.52)$$

The presence of the divergent term u_{31} in the solution (10.50) implies that either there is no periodic solution or we have overlooked an essential aspect of the SHG dynamics. The difficulty can be partially resolved by noting that, because the oscillation amplitude b_1 is still undefined, we can choose b_1 to cancel the real part of u_{31}/b_1

$$|b_1|^2 = 34/99, \qquad a = +1 \qquad (10.53)$$

With this choice for b_1 we are left with $u_{31} = 6ib_1/11$. The fact that u_{31}/b_1 is purely imaginary signs an oscillatory correction to the first order-expression (10.42). Indeed, if instead of $\exp(\pm i\tau)$ the first-order solution were of the form $\exp[\pm i(1 + \sigma)\tau]$ with $\sigma = 6\eta^2/11$, a perturbation expansion of (10.42) would have produced at third order in η the divergent term $6ib_1\tau/11$. There are two ways to interpret this modified exponential. The earliest interpretation was proposed by Poincaré and Lindstedt [5]. They deduced from the divergence that the oscillation frequency cannot be assumed to be constant but that it is η-dependent. Therefore in the expansions (10.40) the time τ must be rescaled, that is, replaced by a new time $\tau' = \omega\tau$ where $\omega = 1 + 6\eta^2/11 + \mathcal{O}(\eta^3)$. The modern approach is to consider that the functions x, y, u, and v depend on the original time τ *and* on a second and independent time, $\mathrm{T} = \eta^2\tau$. This forms the basis of the multiple time scale analysis. The two approaches are equivalent and when applied to an exactly soluble equation, they produce the same solutions. We adopt the second approach. The new time scale is also easy to understand physically. With the definition (10.35) of the small parameter η, the two roots (10.33) that cause the Hopf bifurcation become

$$\lambda_{3,4} = \eta^2 a/3 \pm i(1 - 2\eta^2 a/3 + \cdots) \equiv \gamma(\eta) \pm i\omega(\eta) \qquad (10.54)$$

Hence the oscillations whose frequency is $\omega = 1 + \mathcal{O}(\eta^2)$ are either damped or amplified on a time scale that is proportional to $1/\eta^2$. The characteristic time $1/\gamma$ is determined by the control parameter η. It is therefore not related to the intrinsic dynamics of the system but expresses the critical slowing down in the neighborhood of the critical point. Hence it makes sense to consider that the new time $T = \eta^2 \tau$ is independent of the physical time τ. As a result of this discussion, we replace the too naive expansions (10.40) by the more sophisticated expansions

$$x(\tau, \eta) = \eta^2 x_2(\tau, T) + \mathcal{O}(\eta^4)$$
$$y(\tau, \eta) = \eta^2 y_2(\tau, T) + \mathcal{O}(\eta^4)$$
$$u(\tau, \eta) = \eta u_1(\tau, T) + \eta^3 u_3(\tau, T) + \mathcal{O}(\eta^5)$$
$$v(\tau, \eta) = \eta v_1(\tau, T) + \eta^3 v_3(\tau, T) + \mathcal{O}(\eta^5) \tag{10.55}$$

For differentiation, we use the chain rule

$$z' = dz/d\tau = \partial z/\partial \tau + \eta^2 \, \partial z/\partial T \tag{10.56}$$

for any function z. We insert the expansions (10.55) into the evolution equations (10.36)–(10.39) and use the chain rule (10.56). Canceling the coefficient of each power of η leads again to a sequence of linear problems. The first- and second-order equations remain unaffected so that the solutions (10.42) and (10.44) are modified only by the fact that the unknown functions b_1 and b_2 are functions of the slow time τ. At third order, the equations have the same structure as (10.45)–(10.46). However, the coefficients P_1 and Q_1 become

$$P_1 = b_1[a(1 + i)/3 + |b_1|^2(5i - 20)/17] - \partial b_1/\partial T$$
$$Q_1 = -b_1[2a/3 + 2|b_1|^2(9i - 2)/17] - (i - 1)\,\partial b_1/\partial T \tag{10.57}$$

The condition

$$u_{31} = [(1 + i)P_1 + Q_1]/2i = 0 \tag{10.58}$$

ensures that the secular terms in the third-order solution cancel identically. The novelty is that now this condition leads to a differential equation on the slow time scale

$$\partial b_1/\partial T = b_1[a(1 + i)/3 + |b_1|^2(21i - 33)/34] \tag{10.59}$$

Let us introduce the polar decomposition $b_1 = \rho \exp(i\theta)$ with real ρ and θ. The amplitude satisfies the equation

$$d\rho/d T = \rho(a/3 - \rho^2 33/34) \tag{10.60}$$

Its solution is

$$\rho^2(T) = \frac{2a}{3}\left(\frac{33}{17} + e^{-2aT/3}\left[\frac{2a}{3\rho^2(0)} - \frac{33}{17}\right]\right)^{-1} \tag{10.61}$$

In the long time limit, $\rho^2(T) \to 0$ if $a = -1$ but $\rho^2(T) \to 34/99$ if $a = 1$. The solution $\rho = 0$ corresponds to the steady state that is indeed stable below the bifurcation point, that is, for $a = -1$. Above the bifurcation point ($a = 1$), the stable solution is $\rho^2 = 34/99$. The phase equation is

$$d\theta/d\tau = a/3 + 21\rho^2(\tau)/34 \qquad (10.62)$$

In the long time limit, this gives $\theta(\tau) = 6\tau/11$, which gives the frequency we had guessed in the previous section. Note that the frequency 6/11 is different from, though admittedly close to, the correction 2/3 of the steady state frequency found in (10.54). A number of authors mistakenly assume that close to the Hopf bifurcation the two frequencies are equal.

Thus we see that the multiple time scale analysis provides a differential equation for the first-order amplitude that gives both the long time solution and its stability. This solves the initial value problem. Sometimes, the amplitude equation itself displays an instability. This becomes a secondary bifurcation of the complete solution. For instance, a steady bifurcation of the amplitude means the transition from one periodic solution to another periodic solution. A Hopf bifurcation in the amplitude equations means, in general, a transition from the periodic to a quasi-periodic solution. This, however, requires a higher dimensional problem in order to generate an amplitude equation (usually two coupled amplitude equations) whose linear stability produces a pair of complex conjugate roots.

10.4.3 The solvability condition

The method we have used in the preceding section provides the required answer, namely the amplitude and stability of the periodic solution that is the first-order correction to the steady state solution. To achieve this result, we solved explicitly the second- and third-order problems. The solution of the second order problem was necessary to obtain the expression of the inhomogeneous terms P_j and Q_j in the third-order equations (10.45)–(10.46). However, the derivation of the explicit third-order solution is a waste of time because all we eventually used was a relation between P_1 and Q_1. In this example, the waste has been minimal because the algebra was fairly simple. In general, the explicit solution of the third-order equation is a formidable task that it is better to avoid if possible. The solvability condition is a method that has been devised to avoid the explicit resolution of the third-order problem.

To formulate the solvability condition, let us write the first-, second- and third-order problems in vector notation

$$\partial z_1/\partial \tau = \mathcal{M}_1 z_1 \qquad (10.63)$$

$$\partial z_2/\partial \tau = \mathcal{M}_2 z_2 + \mathcal{N}_2(z_1) \qquad (10.64)$$

$$\partial z_3/\partial \tau = \mathcal{M}_1 z_3 + \mathcal{N}_3(z_1, z_2) \qquad (10.65)$$

where the z_j are 2-dimensional column vectors $z_j = \mathrm{col}(u_j, v_j)$, $\mathcal{M}_j$ are constant 2×2 matrices, and $\mathcal{N}_j$ are 2-dimensional column vectors whose components are nonlinear functions of their argument. Problems arise if the in-homogeneous term $\mathcal{N}_j$ oscillates at a frequency of the homogeneous problem $\partial z_j / \partial \tau = \mathcal{M}_k z_j$. A rigorous proof that the solvability condition is a necessary and sufficient condition is out of place here. We adopt the simpler approach, which is to determine the consequences of the statement that (10.65) has peri-odic solutions of finite amplitude. With suitable rescaling of the time, z_3 can be made 2π-periodic. Let us therefore define a scalar product $\langle a, b \rangle$ and an inner product [a,b] for 2π-periodic functions by

$$[a, b] \equiv \int_0^{2\pi} \langle a, b \rangle d\tau, \qquad \langle a, b \rangle \equiv a_1 b_1 + a_2 b_2 \qquad (10.66)$$

With the scalar product, we define the adjoint $\mathcal{M}^+$ of a matrix $\mathcal{M}$ by

$$\langle a, \mathcal{M}^+ b \rangle \equiv \langle \mathcal{M} a, b \rangle \qquad (10.67)$$

With these elements, we can proceed to derive the solvability condition. Let us take the scalar product of (10.65) with a function w. At this stage, the function w is not defined. It is only constrained by the fact that it yields a finite and nontrivial inner product. Thus we have

$$[(\partial z_3 / \partial \tau), w] = [\mathcal{M}_1 z_3, w] + [\mathcal{N}_3(z_1, z_2), w] \qquad (10.68)$$

Integrating by part and using the definition of the adjoint matrix gives

$$-[z_3, \partial w / \partial \tau] = [z_3, \mathcal{M}_1^+ w] + [\mathcal{N}_3(z_1, z_2), w] \qquad (10.69)$$

In deriving this equation, we have assumed that w is also 2π-periodic, which cancels the boundary contribution in the integration by part. If we now specify the function w to be a 2π-periodic solution of the homogeneous adjoint equation, that is, $\partial w / \partial \tau = -\mathcal{M}_1^+ w$, we are left with the solvability condition

$$[\mathcal{N}_3(z_1, z_2), w] = 0 \qquad (10.70)$$

This result is also known in the theory of differential equations as the *Fredholm alternative theorem* because the solvability condition is tantamount to the state-ment that for each eigenvalue of the matrix $\mathcal{M}_1$ either the homogeneous prob-lem (10.63) has a nontrivial amplitude or the inhomogeneous problem (10.65) has a unique continuous solution for every $\mathcal{N}_3$.

Let us apply this result to our SHG equations. The matrix $\mathcal{M}_1$ and its adjoint are

$$\mathcal{M}_1 = \begin{pmatrix} 1 & 1 \\ -2 & -1 \end{pmatrix}, \qquad \mathcal{M}_1^+ = \begin{pmatrix} 1 & -2 \\ 1 & -1 \end{pmatrix} \qquad (10.71)$$

Both matrices have eigenvalues $\lambda = \pm i$. The eigenvectors of $\mathcal{M}_1^+$ are $w = $ col$[1, (1 \pm i)/2]$. Hence the two solvability conditions for (10.45)–(10.46) are

$$\int_0^{2\pi} e^{\pm i\tau}\left[\left(P_3 e^{3i\tau} + P_1 e^{i\tau} + c.c.\right) + \frac{1 \pm i}{2}\left(Q_3 e^{3i\tau} + Q_1 e^{i\tau} + c.c.\right)\right]d\tau = 0$$

$$(10.72)$$

with P_1 and Q_1 defined by (10.57). In fact, there is only one solvability condition, the second condition being the complex conjugate of the first one. Clearly, the contributions of P_3 and Q_3 will cancel in the integral because of the periodic exponentials that do not compensate. The only nonvanishing contributions to the integral are those for which the product of exponentials equals unity

$$\int_0^{2\pi} e^{-i\tau}\left(P_1 e^{i\tau} + \frac{1-i}{2}Q_1 e^{i\tau}\right)d\tau = \int_0^{2\pi}\left(P_1 + \frac{1-i}{2}Q_1\right)d\tau$$

$$= 2\pi\left(P_1 + \frac{1-i}{2}Q_1\right) = \pi(1-i)\big((1+i)P_1 + Q_1\big) = 0 \qquad (10.73)$$

This is precisely the condition $u_{31} = 0$ that cancels identically the secular terms in the solution (10.50), as should be.

10.5 Extensions

In the third and fourth sections of this chapter we have analyzed in detail the equations

$$\partial\mathcal{E}_1/\partial\tau = \mathcal{E}_1^*\mathcal{E}_2 + \mathcal{E}, \qquad \partial\mathcal{E}_2/\partial\tau = -\mathcal{E}_2 - \mathcal{E}_1^2 \qquad (10.74)$$

near the Hopf bifurcation. Does this analysis provide us with a complete picture of the problem? In fact, there is more to cavity SHG than we have described up to now. Let us briefly explore the missing information.

First, equations (10.74) display one more feature, a hysteresis of the periodic solutions. As the pump parameter $\mathcal{E}$ is increased, all characteristics of the periodic solutions (such as frequency, maximum and minimum amplitude of oscillation, mean value) display a domain of the parameter $\mathcal{E}$ where two periodic solutions coexist. This is shown in Figure 10.1 for $\gamma = 0$. In the same figure, we compare the analytical result derived in the preceding sections and the amplitude of the periodic solution obtained by numerical methods. Extensive simulations of equations (10.74) have not found any other feature of this model in the physical domain of parameters. For the bistable domain, the intermediate branch of solutions is unstable, the other two branches are stable and differ mainly in their frequency content: There are more harmonics of the fundamental frequency for the upper branch (defined for large $\mathcal{E}$) than for the lower branch that emerges from the Hopf bifurcation.

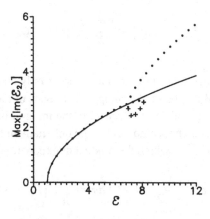

Figure 10.1 Maximum of the imaginary part of $\mathcal{E}_2$ versus the input field $\mathcal{E}$ obtained by integration of equations (10.74). The dots are drawn along the stable periodic solutions, the circles are drawn along the unstable periodic solution and the dashed line is the analytic approximation $2\sqrt{(34/99)(\mathcal{E}-1)}$ obtained in Sections 10.4.2 and 10.4.3 (courtesy of N. Pettiaux and T. Erneux.)

Second, if $\gamma \neq 0$, equations (10.74) are replaced by equations (10.23)–(10.24). Qualitatively, there is no change but all expressions are somewhat more complicated [6]. For instance, the Hopf bifurcation occurs at

$$\mathcal{E}_H = (1+2\gamma)\sqrt{1+\gamma}, \qquad \mathcal{E}_{1,H} = \sqrt{1+\gamma}, \qquad \mathcal{E}_{2,H} = -(1+\gamma) \tag{10.75}$$

and it is characterized by the oscillation frequency $\Omega_H^2 = 1 + 2\gamma$. The eigenvalues $\lambda = \pm i$ of the matrix $\mathcal{M}_1$ and its eigenvectors $e(0) = \mathrm{col}(1, -1+\lambda)$ become

$$\lambda = \pm i\sqrt{1+2\gamma}, \qquad e(\gamma) = \begin{pmatrix} 1 \\ \frac{\lambda-1}{\sqrt{1+\gamma}} \end{pmatrix} \tag{10.76}$$

Finally, in the long time limit the oscillation amplitude is $b_1 = \rho e^{\theta T}$ with

$$|\rho|^2 = \frac{2\sqrt{1+\gamma}[17 + 8\gamma(3+2\gamma)^2]}{(1+2\gamma)[99 + 4\gamma(73 + 56\gamma + 16\gamma^2)]} \tag{10.77}$$

$$\theta = \frac{9(1+2\gamma)(3+2\gamma)}{17 + 8\gamma(3+2\gamma)^2} \frac{|\rho|^2}{\sqrt{1+2\gamma}} \tag{10.78}$$

This solution is stable close to the Hopf bifurcation.

Third, the assumption that has really killed all the complex behavior one expects from nonlinear differential equations like those governing the evolution of SHG is the resonance condition that has been assumed while transforming (10.17)–(10.18) into (10.20)–(10.21). In general, the cavity will not be exactly resonant with the driving field and the second harmonic field. This induces a dispersive correction to the cavity SHG equations which become

$$\partial \mathcal{E}_1/\partial \tau = -(\gamma + i\Delta_1)\mathcal{E}_1 + \mathcal{E}_1^* \mathcal{E}_2 + \mathcal{E}, \qquad \partial \mathcal{E}_2/\partial \tau = -(1 + i\Delta_2)\mathcal{E}_2 - \mathcal{E}_1^2$$
(10.79)

All results obtained for these equations are numerical. The first point to mention is that the Hopf bifurcation is no longer always supercritical: The periodic solution that emerges from the Hopf bifurcation may be unstable [7]. And, of course, chaos is easily found [8] though no systematic investigation of the chaotic domain has been reported.

Finally, let us mention still another reason that justifies the attention devoted to SHG. It was proved in [6] that equations describing other two-photon problems such as two-photon optical bistability and degenerate four-wave mixing reduce, in some limits, to the SHG equations. Thus, the SHG equations may be a reference problem for a class of cavity two-photon problems.

References

[1]   R.W. Boyd, *Nonlinear Optics* (Academic press, New York, 1992).
[2]   P. N. Butcher and D. Cotter, *The Elements of Nonlinear Optics* (Cambridge University Press, 1991).
[3]   A. Yariv and P. Yeh, *Optical Waves in Crystals* (Wiley, New York, 1984).
[4]   P. D. Drummond, K. J. McNeil and D. F. Walls, *Optica Acta* **27** (1980) 321.
[5]   A. H. Nayfeh, *Introduction to Perturbation Techniques* (Wiley, New York, 1981).
[6]   P. Mandel, N. P. Pettiaux, K. Wang, P. Galatola, and L. A. Lugiato, *Phys. Rev. A* **43** (1991) 424.
[7]   V. Zenhlé and P. Mandel, *Opt. Commun.* **66** (1988) 216.
[8]   C. M. Savage and D. F. Walls, *Optica Acta* **30** (1983) 557.

11

Saturable absorbers

11.1 Introduction

Until now, we have always assumed that the cavity losses are linear, that is, field-independent. This need not always be true, and in this chapter we investigate how nonlinear losses affect the operation of a nonlinear optical device. We consider two examples. The first one is the laser with a saturable absorber, the second is parametric amplification in the presence of a saturable absorber.

The modeling of lasers with a saturable absorber (LSA) has a history that is practically as long as that of the laser. Problems related to the LSA are still a subject of debate. An LSA is a laser that contains both an active (or amplifying) medium and a passive (or absorbing) medium. For instance, if population excitation is produced in the laser cavity but inversion is not achieved in the whole cavity, some domains of the laser amplify the radiation and others absorb it. The salient feature of this situation is that *both* amplification and absorption result from the resonant interaction of light with atoms. Hence, both processes contribute to the nonlinear response. One way to look at an LSA is to consider it as a generalization of a laser in which the losses are as nonlinear (i.e., intensity-dependent) as the gain. From the viewpoint of dynamical systems, the LSA is a prototype of competition between nonlinear gain and nonlinear losses. As a result, there has been over the years an irresistible temptation to attribute to saturable absorption properties that are not explained by the standard laser equations, derived in Chapter 1. This is the reason for the huge literature on the LSA and the fact that in many cases the papers turn out, with hindsight, to be of little relevance.

A historical presentation of the LSA can be found in [1] and [2]. In particular, optical bistability resulting from the presence of a saturable absorber was predicted as early as 1964 [3] and verified experimentally the following year. Intracavity spectroscopy, in which the sample to be analyzed is placed inside the lasing cavity to benefit from the much larger intracavity field intensity, is another way to realize an LSA. In the early days of intracavity spectroscopy [4], an inverted Lamb dip was found in gas LSA. It has provided a powerful means of laser stabilization. However, the behavior with which we will be mostly concerned in this chapter is passive Q-switching. The method of Q-switching was proposed by Hellwarth [5] and was a transposition of techniques applied to nuclear magnetic resonance in the microwave domain. The idea is to induce a sudden variation in the quality factor $Q = \omega_c/2\gamma_c$ to induce a giant pulse. In the proposition of Hellwarth, this variation was created by an optical shutter controlled externally. However, the variation of the Q factor can be controlled by either an active or a passive mechanism. The classic example of an active control is a Kerr cell that is inserted in the cavity and driven by a modulated voltage. The passive mechanism of choice is an intracavity saturable medium in which population inversion is not realized [6]. In that case, the absorber is opaque at low intensity and bleached at high intensity.

Let us consider for simplicity the situation where the saturable absorber is placed between the active medium and one of the cavity mirrors. In this configuration, the absorbing medium modifies the cavity losses that become proportional to $\gamma_c^0 + \alpha/(1 + I/I_{sat})$, where γ_c^0 is the empty cavity damping rate, $\alpha > 0$ is the linear loss coefficient of the absorbing medium, I_{sat} is its saturation intensity, and I the intracavity field intensity. For low Q ($I \ll I_{sat}$ and γ_c is maximum), the intracavity field builds up because the light cannot escape the cavity since the absorber is opaque. However, when the intensity exceeds the saturation intensity of the absorbing medium, the absorber losses become proportional to $\gamma_c^0 + \alpha I_{sat}/I$ and sharply decrease: The absorber is bleached by the intracavity field. Hence, light is emitted by the high-Q cavity, and the intracavity field drops below I_{sat} leading to a significant increase of the losses and the end of the light emission until a new cycle can begin, when the field inside the cavity has been able to increase sufficiently to bleach again the absorber. This process is called *passive Q-switching*. It has been one of the first methods used to generate short pulses because the pulse width and repetition rates are directly determined by the atomic response times, which turn out to be quite favorable to the generation of short pulses. From the theoretical viewpoint, passive Q-switching offers a challenge because we have to describe a pulsed solution made of two pieces: Short and intense pulses separated by domains of exponentially small intensity. It has been shown that passive Q-switching can be analyzed by using the method of matched asymptotic expansions [9]. However, this analysis is still quite involved and does not lead to simple results. Therefore we postpone the detailed analysis of passive Q-switching to the DOPOSA, studied in Section 11.3. In both cases, the saturable absorber is modeled by a set of homogeneously broadened two-level atoms.

11.2 Laser with a saturable absorber

In the standard configurations of an LSA, the two media are either separated in the cavity or they share the same volume. For a ring cavity in the uniform field limit, there is no difference in the equations that describe these two problems. Therefore, we can easily derive the evolution equations by considering that the effects of the two media are additive: The contribution to the total atomic polarization is the sum of the active and of the passive atomic polarizations. However, we cannot directly use equations (1.48)–(1.50) because the field and the material variables have already been scaled by means of the amplifying medium parameters. The simplest and surest way to proceed is to begin with the totally unscaled equations (1.29) for the field and (1.26)–(1.27) for the medium. Assuming a single-mode laser and using the additivity of the atomic polarization leads in a straightforward way to

$$dE_0/dt = -\gamma_c E_0 - (N\omega_c \mu/2\varepsilon_0)P_0 - (\overline{N}\omega_c \overline{\mu}/2\overline{\varepsilon}_0)\overline{P}_0 \qquad (11.1)$$

$$dP_0/dt = -[\gamma_\perp + i(\omega_a - \omega_c)]P_0 - (\mu/\hbar)E_0 D \qquad (11.2)$$

$$dD/dt = -\gamma_\parallel(D - D_a) + (\mu/2\hbar)(P_0^* E_0 + P_0 E_0^*) \qquad (11.3)$$

$$d\overline{P}_0/dt = -[\overline{\gamma}_\perp + i(\overline{\omega}_a - \omega_c)]\overline{P}_0 - (\overline{\mu}/\hbar)E_0\overline{D} \qquad (11.4)$$

$$d\overline{D}/dt = -\overline{\gamma}_\parallel(\overline{D} - \overline{D}_a) + (\overline{\mu}/2\hbar)(\overline{P}_0^* E_0 + \overline{P}_0 E_0^*) \qquad (11.5)$$

The convention in this domain is that the overbar affects the variables of the absorbing medium. The difference between the absorbing and amplifying media resides in the value of the population difference: $D_a > 0$ signs an inversion of population, whereas $\overline{D}_a < 0$ indicates that the upper level is less populated than the lower level.

Equations (11.1)–(11.5) describe the Maxwell–Bloch variables in a reference frame that rotates with frequency ω_c. To study the steady states, it is more sensible to shift to a reference frame that rotates with the lasing frequency Ω

$$dE_0/dt = -[\gamma_c + i(\omega_c - \Omega)]E_0 - (N\omega_c\mu/2\varepsilon_0)P_0 - (\overline{N}\omega_c\overline{\mu}/2\overline{\varepsilon}_0)\overline{P}_0 \quad (11.6)$$

$$dP_0/dt = -[\gamma_\perp + i(\omega_a - \Omega)]P_0 - (\mu/\hbar)E_0 D \qquad (11.7)$$

$$dD/dt = -\gamma_\parallel(D - D_a) + (\mu/2\hbar)(P_0^* E_0 + P_0 E_0^*) \qquad (11.8)$$

$$d\overline{P}_0/dt = -[\overline{\gamma}_\perp + i(\overline{\omega}_a - \Omega)]\overline{P}_0 - (\overline{\mu}/\hbar)E_0\overline{D} \qquad (11.9)$$

$$d\overline{D}/dt = -\overline{\gamma}_\parallel(\overline{D} - \overline{D}_a) + (\overline{\mu}/2\hbar)(\overline{P}_0^* E_0 + \overline{P}_0 E_0^*) \qquad (11.10)$$

In that case the steady state equations for the LSA are obtained by setting the time-derivatives on the left-hand side of equations (11.6)–(11.10) equal to zero. Solving for the material equations gives a complex steady state field equation

$$E_0 \left[\gamma_c(1 + i\Delta) - \frac{N\omega_c\mu^2 D_a(1 - i\delta)}{2\varepsilon_o\hbar\gamma_\perp \left(1 + \delta^2 + \frac{\mu^2|E_o|^2}{\hbar^2\gamma_\perp\gamma_\parallel} \right)} \right.$$

$$\left. \frac{\overline{N}\omega_c\overline{\mu}^2\overline{D}_a(1 - i\overline{\delta})}{2\overline{\varepsilon}_o\hbar\overline{\gamma}_\perp \left(1 + \overline{\delta}^2 + \frac{\overline{\mu}^2|E_o|^2}{\hbar^2\overline{\gamma}_\perp\overline{\gamma}_\parallel} \right)} - \right] = 0 \qquad (11.11)$$

where we have used the usual definitions for the detunings

$$\Delta = (\omega_c - \Omega)/\gamma_c, \qquad \delta = (\omega_a - \Omega)/\gamma_\perp, \qquad \overline{\delta} = (\overline{\omega}_a - \Omega)/\overline{\gamma}_\perp \qquad (11.12)$$

We introduce the resonant pump parameters $A > 1$ for the amplifying medium and $\overline{A} < 0$ for the absorbing medium

$$A = N\omega_c\mu^2 D_a/2\hbar\varepsilon_0\gamma_\perp\gamma_c, \qquad \overline{A} = \overline{N}\omega_c\overline{\mu}^2\overline{D}_a/2\hbar\overline{\varepsilon}_o\overline{\gamma}_\perp\gamma_c \qquad (11.13)$$

and two ratio parameters

$$a = \overline{\gamma}_\perp\overline{\gamma}_\parallel\mu^2/(\gamma_\perp\gamma_\parallel\overline{\mu}^2), \qquad b = (\gamma_\perp/\overline{\gamma}_\perp)^2 \qquad (11.14)$$

The parameter a measures the relative saturability of the two media. It is an important parameter whose magnitude determines many properties of the LSA. With these definitions, the complex steady state equation (11.11) yields the two real equations

$$E_s \left(1 + \frac{A}{1 + \delta^2 + I} + \frac{\overline{A}}{1 + \overline{\delta}^2 + aI} \right) = 0 \qquad (11.15)$$

$$E_s \left(\Delta + \frac{A\delta}{1 + \delta^2 + I} + \frac{\overline{A}\overline{\delta}}{1 + \overline{\delta}^2 + aI} \right) = 0 \qquad (11.16)$$

The intensity I that appears in these two equations has the usual definition $I = \mu^2|E_o|^2/(\hbar^2\gamma_\perp\gamma_\parallel) \equiv |E_o|^2/I_s$, where I_s is the saturation intensity of the active medium. The first equation generalizes the "state equation" that determines the field intensity, whereas the second equation generalizes the dispersion relation that determines the field oscillation frequency. To simplify the analysis, we assume that both atomic frequencies and the cavity frequency are equal: $\omega_c = \omega_a = \overline{\omega}_a$. Introducing the dimensionless parameters

$$d = \gamma_\perp/\gamma_c, \qquad \overline{d} = \overline{\gamma}_\perp/\gamma_c \qquad (11.17)$$

yields the simpler state equation for the field amplitude and dispersion relation

$$E_s\left(1 - \frac{A}{1 + \delta^2 + I} - \frac{\overline{A}}{1 + b\delta^2 + aI}\right) = 0 \qquad (11.18)$$

$$E_s(\Omega - \omega_c)\left(1 + \frac{A/d}{1 + \delta^2 + I} + \frac{\overline{A}/\overline{d}}{1 + b\delta^2 + aI}\right) = 0 \qquad (11.19)$$

Despite the simplification introduced by the degenerate tuning assumption $\omega_c = \omega_a = \overline{\omega}_a$, the steady state equations (11.18)–(11.19) offer an exceptional richness of solutions. Solving these equations for A and $\overline{A}$ indicates that to have solutions with $A > 0$ and $\overline{A} < 0$ requires $b > 1$. Three classes of solutions can be found.

1. The trivial solution $E_s = 0$, corresponding to the laser being off
2. The resonant solutions $\Omega = \omega_c$ with a finite intensity

$$I_\pm = \frac{1}{2a}\left(a(A - 1) - 1 + \overline{A} \pm \sqrt{[a(A - 1) - 1 + \overline{A}]^2 - 4a(1 - A - \overline{A})}\right)$$

$$(11.20)$$

3. The off-resonant solutions with a finite intensity

$$\tilde{I} = \frac{1}{a - b}\left[b - 1 + (\overline{d} - d)\left(\frac{\overline{A}}{\overline{d}(d + 1)} + \frac{Ab}{d(\overline{d} + 1)}\right)\right] \qquad (11.21)$$

$$\delta = \left\{\frac{1}{b - a}\left[a - 1 + (\overline{d} - d)\left(\frac{\overline{A}}{\overline{d}(d + 1)} + \frac{Aa}{d(\overline{d} + 1)}\right)\right]\right\}^{1/2} \qquad (11.22)$$

For the resonant solutions, the classification of the steady states is easy to obtain, based on the condition that the intensity (11.20) must be real. This requires the expression under the square root to be positive: $(aA - X_-)(aA - X_+) > 0$ with $X_\pm = a - 1 - \overline{A} \pm 2[-(a - 1)\overline{A}]^{1/2}$. The two possibilities are the following:

1. $0 < a < 1 - 1/\overline{A}$. The threshold is at $A_{th} = 1 - \overline{A}$. If $A < A_{th}$, the only solution is the trivial solution $I = 0$. Above threshold, $A > A_{th}$, the two solutions $I = 0, I_+$ coexist.
2. $a > 1 - 1/\overline{A}$. Three domains have to be considered. Let us define the pump parameter that corresponds to a limit point

$$B_\ell = \frac{1}{a}\{a - 1 - \overline{A} + 2[\overline{A}(1 - a)]^{1/2}\} \qquad (11.23)$$

If $A < B_\ell$, there is only the trivial solution $I = 0$. If $B_\ell < A < 1 - \overline{A}$,

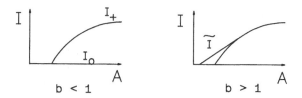

Figure 11.1 Intensity versus amplifying medium pump parameter: steady state solutions for $1 < a < 1 - 1/\overline{A}$. After [1].

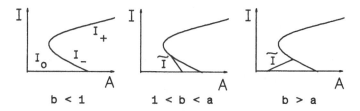

Figure 11.2 Intensity versus amplifying medium pump parameter: steady state solutions for $a > 1 - 1/\overline{A}$ and $\overline{A} > (1 - a)(1 - 1/\overline{A}_c)^2$. After [1].

all three solutions $I = 0, I_+$ and I_- coexist. Finally, if $A > 1 - \overline{A}$, only $I = 0, I_+$ coexist.

The nonresonant solution emerges from the trivial solution at

$$\tilde{A}_0 = (1 - \overline{A})\frac{d(\overline{d} + 1)}{d(d + 1)} + \frac{\overline{d} + 1}{d + 1}(1 + d + \overline{d}) \qquad (11.24)$$

In the plane (I, A), the nonresonant solution is a straight line reaching the finite intensity solution at $A = \tilde{A}_1$ obtained by setting $\delta = 0$ in (11.22). This gives

$$\tilde{A}_1 = \frac{d(\overline{d} + 1)}{a}\left(\frac{1 - a}{\overline{d} - d} - \frac{\overline{A}}{\overline{d}(d + 1)}\right) \qquad (11.25)$$

If $1 < a < b$, the nonresonant intensity has a negative slope in the (I, A) plane; that is, $\tilde{A}_0 < \tilde{A}_1$. On the contrary, if $1 < b < a$, the slope of the nonresonant solution is positive and $\tilde{A}_0 > \tilde{A}_1$. Furthermore, we can define another critical parameter by

$$\overline{A}_c = (1 - a)\left(\frac{\overline{d}(d + 1)}{\overline{d} - d}\right)^2 \qquad (11.26)$$

If $\overline{A} > \overline{A}_c$, the nonresonant solution reaches the I_- branch, whereas if $\overline{A} < \overline{A}_c$,

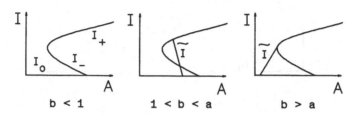

Figure 11.3 Intensity versus amplifying medium pump parameter: steady state solutions for $a > 1 - 1/\overline{A}$ and $\overline{A} < (1 - a)(1 - 1/\overline{A}_c)^2$. After [1].

the nonresonant solution reaches the I_+ branch. The various solutions are shown in Figures 11.1 to 11.3.

The occurrence of the nonresonant solutions is a feature that has been found via the same mechanism as described here in other systems characterized by a competition between nonlinear gain and nonlinear loss. The emergence of these solutions means that the nonlinear losses are able to induce a dispersive response ($\delta \neq 0$) of a resonant system ($\omega_c = \omega_a = \overline{\omega}_a$).

11.3 Parametric oscillator with a saturable absorber

11.3.1 Formulation and steady states

In the preceding section, both the amplifying and the absorbing media supported single-photon transitions. A variant on this theme is the intracavity optical parametric amplifier with a saturable absorber that combines two-photon and one-photon processes.

We have seen in Chapter 10 that two-photon transitions are possible in media that lack the reflexion symmetry. Let us first describe the optical parametric oscillator without absorber:

- the cavity is driven by a coherent external field of frequency 2ω
- the amplifying medium absorbs a photon of the driving field and emits two photons at the same frequency ω
- the phase matching $k^{(2\omega)} = 2k^{(\omega)}$ is verified

This characterizes a degenerate optical parametric oscillator (DOPO) because the two photons emitted have the same frequency. In the nondegenerate optical parametric oscillator, the two emitted photons have different frequencies, ω_1 and ω_2, such that their sum equals 2ω and the phase matching condition becomes $\mathbf{k}^{(\omega)} = \mathbf{k}^{(\omega_1)} + \mathbf{k}^{(\omega_2)}$. It follows from the derivation presented in Section 10.2 that the DOPO is described by the evolution equations

$$\partial E_1/\partial t = -\gamma_{c1}E_1 + iaE_2E_1^* \tag{11.27}$$

$$\partial E_2/\partial t = -\gamma_{c2}E_2 + ibE_1^2 + E_{\text{ext}} \tag{11.28}$$

Now we consider a DOPO whose cavity also contains a saturable absorber that absorbs photons at frequency ω. We model the absorber by a two-level medium, as in the preceding section. Therefore, to describe a DOPO with a saturable absorber (DOPOSA), we combine equations (11.4)–(11.5) for the saturable absorber with equations (11.27)–(11.28) for the DOPO. We couple the two sets of equations by adding to the equation describing the evolution of E_1 the contribution of the atomic polarization originating in the passive medium. This leads to

$$dE_1/dt = -[\gamma_{c1} + i(\nu_1 - \omega)]E_1 - (\overline{N}\omega_c\overline{\mu}/2\overline{\varepsilon}_0)\overline{P}_0 + iaE_2E_1^* \tag{11.29}$$

$$dE_2/dt = -[\gamma_{c2} + i(\nu_2 - 2\omega)]E_2 + ibE_1^2 + E_{\text{ext}} \tag{11.30}$$

$$d\overline{P}_o/dt = -[\overline{\gamma}_\perp + i(\overline{\omega}_a - \omega)]\overline{P}_o - (\overline{\mu}/\hbar)E_1\overline{D} \tag{11.31}$$

$$d\overline{D}/dt = -\overline{\gamma}_\parallel(\overline{D} + |\overline{D}_a|) + (\overline{\mu}/2\hbar)(\overline{P}_o^*E_1 + \overline{P}_oE_1^*) \tag{11.32}$$

The new parameters are ν_1 and ν_2, the cavity frequencies closest to ω and 2ω, respectively. A change of variables will reduce the number of relevant parameters. We introduce the new variables

$$E_1 = \mathcal{E}_1\gamma_{c1}/\sqrt{ab}, \qquad E_2 = -i\mathcal{E}_2\gamma_{c1}/a, \qquad \overline{P}_0 = \mathcal{P}\overline{\mu}|\overline{D}_a|/\hbar\sqrt{ab},$$

$$\overline{D} = -D|\overline{D}_a|, \qquad E_{\text{ext}} = -i\mathcal{E}\gamma_{c1}^2/a, \qquad \tau = \gamma_{c1}t,$$

$$\Delta_j = (\nu_j - j\omega)/\gamma_{c1}, \qquad \Delta_a = (\overline{\omega}_a - \omega)/\gamma_{c1}, \qquad \gamma = \gamma_{c2}/\gamma_{c1},$$

$$\gamma_1 = \overline{\gamma}_\perp/\gamma_{c1}, \qquad \gamma_2 = \overline{\gamma}_\parallel/\gamma_{c1} \tag{11.33}$$

Two new physical parameters are

$$R = \frac{\overline{N}\omega_c\overline{\mu}^2|\overline{D}_a|}{2\hbar\overline{\varepsilon}_0\gamma_{c1}\overline{\gamma}_\perp}, \qquad S = \frac{\overline{\mu}^2\gamma_{c1}\gamma_{c2}}{\hbar^2 ab\overline{\gamma}_\perp\overline{\gamma}_\parallel} \tag{11.34}$$

The parameter R is the pump parameter of the absorbing medium, whereas S is a measure of the saturation intensity for mode 1. In the new variables, the evolution equations of the DOPOSA are

$$d\mathcal{E}_1/d\tau = -(1 + i\Delta_1)\mathcal{E}_1 + \mathcal{E}_1^*\mathcal{E}_2 - R\gamma_1\mathcal{P} \tag{11.35}$$

$$d\mathcal{E}_2/d\tau = -(\gamma + i\Delta_2)\mathcal{E}_2 - \mathcal{E}_1^2 + \mathcal{E} \tag{11.36}$$

$$d\mathcal{P}/d\tau = -(\gamma_1 + i\Delta_a)\mathcal{P} + \mathcal{E}_1 D \tag{11.37}$$

$$dD/d\tau = -\gamma_2(D - 1) - (\gamma_1\gamma_2 S/2\gamma)(\mathcal{P}^*\mathcal{E}_1 + \mathcal{E}_1^*\mathcal{P}) \tag{11.38}$$

To simplify the discussion, we assume that the condition of perfect resonance is satisfied; that is, $\Delta_1 = \Delta_2 = \Delta_a = 0$. In that case, nothing is lost for the description of the steady states by assuming real solutions. There are two steady states. The trivial solution

$$\mathcal{E}_1 = \mathcal{P} = 0, \qquad D = 1, \qquad \mathcal{E}_2 = \mathcal{E}/\gamma \qquad (11.39)$$

describes the DOPOSA below the lasing threshold. The lasing solution is given by

$$D = 1/(1 + SI_1), \qquad \mathcal{P} = \mathcal{E}_1/[\gamma_1(1 + SI_1)],$$
$$\mathcal{E}_2 = \mathcal{E}/\gamma - I_1, \qquad \mathcal{E} = \gamma[1 + I_1 + R/(1 + SI_1)] \qquad (11.40)$$

with the definition $I_1 = \mathcal{E}_1^2/\gamma$. We will concentrate our analysis on the lasing solution (11.40). A report on the properties of the nonlasing solution can be found in [11]. The lasing and the nonlasing solutions coincide at $\mathcal{E} \equiv \mathcal{E}_{th} = \gamma(1 + R)$. If $\mathcal{E} > \mathcal{E}_{th}$, the nonlasing solution is unstable.

Even if there is no saturable absorber in the cavity, there is a threshold $\mathcal{E}_{th}^0 = \gamma$ for the emergence of a finite field amplitude $\mathcal{E}_1$. This makes DOPO and SHG radically different processes because SHG is thresholdless.

11.3.2 Small amplitude periodic solutions

To concentrate on the essentials, we assume that the saturable absorber relaxes much faster than the cavity fields. In that case, we may adiabatically eliminate the two material variables $\mathcal{P}$ and D, which leads to a pair of field equations

$$dI_1/d\tau = 2I_1\left(\mathcal{E}_2 - 1 - \frac{R}{1 + SI_1}\right), \qquad d\mathcal{E}_2/d\tau = \gamma(\mathcal{E} - \mathcal{E}_2 - I_1) \quad (11.41)$$

The coupling between $\mathcal{E}_1^2$ and $\mathcal{E}_2$ results quite naturally from the fact that $\mathcal{E}_1$ is associated with photons of frequency ω whereas $\mathcal{E}_2$ is associated with photons of frequency 2ω.

The steady state solution (11.40) is monostable if $RS < 1$ and bistable if $RS > 1$. In the latter case, there is a limit point at

$$\mathcal{E} = \mathcal{E}_{lim} = \gamma(1 - 1/S + 2\sqrt{R/S}) \qquad (11.42)$$

Using the dynamical equations (11.41), a linear stability analysis of the steady state solution (11.40) yields a quadratic equation

$$\lambda^2 + a_1\lambda + a_2 = 0 \qquad (11.43)$$

$$a_1 = \gamma - 2I_1 RS/(1 + SI_1)^2, \qquad a_2 = 2I_1\gamma[1 - RS/(1 + SI_1)^2]$$

If $RS > 1$, the lower branch of the steady solution is unstable because $a_2 < 0$. More interesting is the existence of a domain defined by $a_1 = 0$ and $R > 2\gamma$

in which the steady solution is also unstable. This domain is bounded by Hopf bifurcations located at the zeros of a_1

$$I_1 \equiv I_{H,\pm} = \{R - \gamma \pm \sqrt{R(R - 2\gamma)}\}/\gamma S \qquad (11.44)$$

At these bifurcations we have

$$\lambda_H = \pm i\omega, \qquad \omega^2 = \gamma(2I_H - \gamma) \qquad (11.45)$$

We can list the various situations in terms of the parameter

$$S_\pm = 2\{R - \gamma \pm \sqrt{R(R - 2\gamma)}\}/\gamma^2 \qquad (11.46)$$

1. In the monostable case, $RS < 1$, the two Hopf bifurcations exist on the branch $I_1 > 0$.
2. In the bistable case, $RS > 1$, the following classification is obtained.

   - For $S_- < S_+ < S$, there is no Hopf bifurcation on the upper branch that is stable against infinitesimal perturbations.
   - For $S_- < S < S_+$, there is one Hopf bifurcation on the upper branch.
   - For $S = S_-$, one Hopf bifurcation coincides with the limit point at $\mathcal{E}_{\lim}$; the second Hopf bifurcation occurs on the upper branch for higher values of the pump parameter A.
   - For $1/R < S < S_-$, there are two Hopf bifurcations on the upper branch.

We can study the stability of the periodic solutions that emerge from the Hopf bifurcations with the method devised in Chapter 10. These are the small amplitude periodic solutions of the problem. To determine their amplitude and stability, we define a vicinity of the Hopf bifurcation by

$$\mathcal{E} = \mathcal{E}_H + a\eta^2, \qquad a = \pm 1 \qquad (11.47)$$

where $\mathcal{E}_H = 1 + R/(1 + SI_H) + I_H$. These definitions hold for either Hopf bifurcation. For reasons explained in Chapter 10, we also introduce two times

$$T = \omega\tau, \qquad \mathrm{T} = \eta^2\tau \qquad (11.48)$$

in terms of which we seek solutions of the evolution equations (11.41) in the form of power series

$$\mathcal{E}_2 = \mathcal{E}_{2,H} + \eta x_1 + \mathcal{O}(\eta^2), \qquad I_1(\tau, \eta) = I_H + \eta y_1 + \mathcal{O}(\eta^2) \quad (11.49)$$

The first-order equations are

$$\omega \, \partial x_1/\partial T = -\gamma(x_1 + y_1), \qquad \omega \, \partial y_1/\partial T = 2I_H[x_1 + RSy_1/(1 + SI_H)^2] \qquad (11.50)$$

and their solution is

$$\begin{pmatrix} x_1 \\ y_1 \end{pmatrix} = \alpha(\tau)\begin{pmatrix} p \\ 1 \end{pmatrix}\exp(i\mathrm{T}) + c.c., \qquad p = -\gamma/(\gamma + i\omega) \qquad (11.51)$$

The function $\alpha(\tau)$ is still unknown. As in Chapter 10, and for the same reasons, we have to derive a solvability condition that arises in the third-order equations to obtain an equation that determines $\alpha(\tau)$. We skip the technical steps that do not bring any relevant information and quote the result, that is, the amplitude equation

$$d|\alpha|/d\tau = (P_\alpha + Q_\alpha|\alpha|^2)|\alpha| \qquad (11.52)$$

$$P_\alpha = \frac{\gamma(1 - SI_H)a}{(2I_H - \gamma)(1 + SI_H)}, \qquad Q_\alpha = \frac{\gamma(2S^2I_H^2 - 4SI_H + \gamma S)}{2I_H(2I_H - \gamma)(1 + SI_H)^2}$$

The trivial solution, $|\alpha| = 0$, corresponds to the steady state (11.40). The nontrivial solution $|\alpha|^2 = -P_\alpha/Q_\alpha > 0$ corresponds to the small amplitude periodic solution. The condition $P_\alpha/Q_\alpha < 0$ determines the sign of a and therefore the direction of the bifurcation. It also determines the stability of the periodic solution. Six cases have been identified; they are given in Figure 11.4. If there are two Hopf bifurcations, they are connected by a continuous branch of periodic solutions. If there is only one Hopf bifurcation, the branch of periodic solutions terminates with a limit point.

11.3.3 Passive Q-switching

The real treat when studying the good cavity DOPOSA described by equations (11.41) is the occurrence of passive Q-switching (PQS) in the limit $\gamma = \gamma_{c2}/\gamma_{c1} \to 0$. Figures 11.5 and 11.6 display a clear example of such a PQS. In Figure 11.7 we show the bifurcation diagram for the same values of R, S, and γ.

The analysis of the previous section is obviously unable to cope with the phenomenon of PQS. A method, based on singular perturbation methods and matched asymptotic expansions has been used to analyze PQS in the Lorenz equations, [i.e., the laser equations (1.58)–(1.60) on resonance] [10] and in the LSA [9]. The same method is applied to the DOPOSA in this section. It is based on the observation that solutions like those displayed in Figure 11.5 can be decomposed in two parts. During a short period of time, the pulse takes place. If γ increases, the peak intensity increases and the pulse width decreases. This part of the pulse corresponds to the portion of the curve from Q to P in the phase space represented in Figure 11.6. Between consecutive pulses, I_1 is vanishingly small and $\mathcal{E}_2$ builds up until it reaches its maximum. This second domain of the pulsed solution corresponds to the portion of the curve in the phase plane that

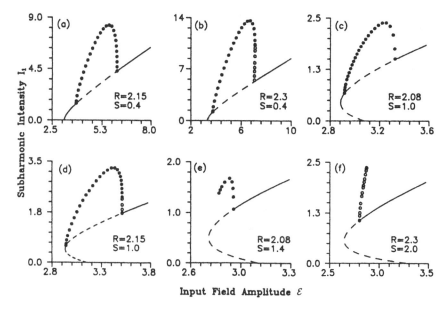

Figure 11.4 Classification of the bifurcation diagrams of the DOPOSA [equation (11.41)] in the good cavity limit, with $\gamma = 1$, in terms of the parameter $S = 2\{\gamma^2 - [R - 2\gamma + \sqrt{R^2 - 2\gamma R}]^2\}/\gamma^3$, and the constants $\beta_1 \simeq 2.20574$ and $\beta_2 \simeq 2.11803$. Solid lines are stable steady states, dashed lines are unstable steady states, dots are stable periodic solutions, and circles are unstable periodic solutions. For the periodic solutions, the maximum amplitude is plotted. (a) $2\gamma < R < \gamma\beta_1$ and $SR < 1$. (b) $R > \gamma\beta_1$ and $SR < 1$. (c) either $2\gamma < R < \gamma\beta_1$ and $1 < RS < S_-$ or $\gamma\beta_2 < R < \gamma\beta_1$ and $1 < RS < RS$. (d) either $\gamma\beta_2 < R < \gamma\beta_1$ and $S < S < S_-$ or $R > \gamma\beta_1$ and $1 < RS < RS_-$. (e) $2\gamma < R < \gamma\beta_2$ and $S_- < S < S$. (f) either $R > \gamma\beta_2$ and $S_- < S < S_+$ or $2\gamma > R > \gamma\beta_2$ and $S < S < S_+$. Reproduced from [11].

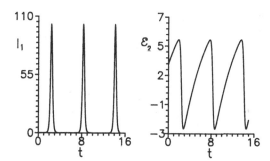

Figure 11.5 Pulsed solutions for the fields I_1 and $\mathcal{E}_2$ versus time for $R = S = 1$, $\mathcal{E} = 10$, and $\gamma = 0.2$. The two Hopf bifurcations are located at $\mathcal{E}_{H1} \simeq 2.014$ and $\mathcal{E}_{H2} \simeq 8.986$. After [11].

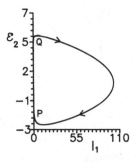

Figure 11.6 Pulsed solutions in the phase plane $(I_1, \mathcal{E}_2)$ for the same parameters as in Figure 11.5. The arrows indicate in which direction the curve is followed as time increases. After [11].

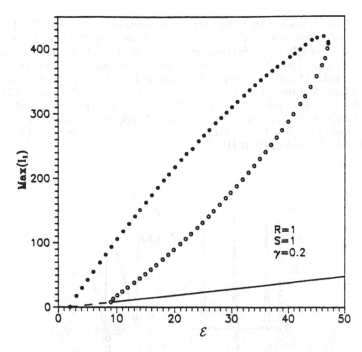

Figure 11.7 Bifurcation diagram of the periodic solutions of equation (11.41). Same drawing conventions as in Figure 11.4. Note the difference in amplitude with Figure 11.4(b) for $\gamma = 1$. Reproduced from [11].

goes from P to Q. This suggests that we seek an approximate solution in each of the two domains and match these two solutions. Usually, it is not difficult to find these approximate solutions. The problem is the matching at the domain boundaries. In the case of the Lorenz equations and the LSA, the matching leads to rather complicated equations that must be analyzed asymptotically. On the contrary, in the DOPOSA, the results take a simple form. Another situation where the method of matched asymptotic expansions was used successfully is in Section 3.1.1, where three solutions were matched sequentially.

The outer solution: Between consecutive pulses (from P to Q in Figure 11.6), $I_1 \ll 1$ and $\mathcal{E}_2 = \mathcal{O}(1)$. Furthermore the period of the pulses increases if γ decreases. Let us therefore seek a power series solution of the form

$$I_1(\tau, \gamma) = \gamma J_1(\tau, \sigma) + \gamma^2 J_2(\tau, \sigma) + \mathcal{O}(\gamma^3),$$
$$\mathcal{E}_2(\tau, \gamma) = B_0(\tau, \sigma) + \gamma B_1(\tau, \sigma) + \mathcal{O}(\gamma^2) \tag{11.53}$$

where we have introduced the new time scale $\sigma = \gamma \tau$. Inserting these series into the coupled mode equations (11.41) leads to a sequence of equations. The first equations are

$$\partial B_0/\partial \tau = 0 \tag{11.54}$$

$$\partial B_1/\partial \tau = \mathcal{E} - B_0 - \partial B_0/\partial \sigma \tag{11.55}$$

$$\partial J_1/\partial \tau = 2J_1(B_0 - 1 - R) \tag{11.56}$$

From (11.54) we obtain $B_0 = B_0(\sigma)$. With this result, it follows from (11.55) that $\mathcal{E} - B_0 - \partial B_0/\partial \sigma = 0$ and therefore

$$B_0(\sigma) = \mathcal{E} + C_1 e^{-\sigma} \tag{11.57}$$

where C_1 is an integration constant to be determined. Having found $B_0(\sigma)$, it is easy to integrate (11.56) to obtain the dominant contribution to the subharmonic intensity

$$J_1(\sigma) = C_2 \exp\{2[(\mathcal{E} - 1 - R)\sigma - C_1(\exp(\ \sigma) - 1)]/\gamma\} \tag{11.58}$$

To obtain this result, we have expressed the solution B_0 as $\mathcal{E} + C_1 \exp(-\gamma \tau)$ in (11.56). Direct integration of this equation yields $J_1(\tau) = C_2 \exp\{2(\mathcal{E} - 1 - R)\tau - 2C_1(\exp(-\gamma \tau) - 1)/\gamma\}$, which is equivalent to (11.58). In this way, we retain in J_1 the evolution on the two time scales τ and σ. The constant C_2 is determined, for instance, by requiring that $\sigma = 0$ at the point P in Figure 11.6: $J_1(0) = C_2$. The solution (11.58) converges until the point Q is reached, where the pulse begins to emerge. At the point Q, the time is equal to the period p minus the pulse width. Since the pulse width is vanishingly small as $\gamma \to 0$, the time at Q is a good approximation of the period p. However, at Q the intensity is still negligible. Hence, we obtain a first equation for the pulse period by setting $J_1(\sigma)$ at Q equal to $J_1(p)$, which is also equal to $J_1(0)$

$$(\mathcal{E} - 1 - R)p + C_1[1 - \exp(-p)] = 0 \qquad (11.59)$$

The inner solution: The inner solution refers to the description of the pulse itself. Since I_1 increases with decreasing γ, we seek a solution of the evolution equations (11.41) in the form of the power series

$$\begin{aligned} I_1(\tau, \gamma) &= \gamma^{-1}K_0(\tau) + K_1(\tau) + \mathcal{O}(\gamma), \\ \mathcal{E}_2(\tau, \gamma) &= G_0(\tau) + \gamma G_1(\tau) + \mathcal{O}(\gamma^2) \end{aligned} \qquad (11.60)$$

The dominant order leads to a conservative system

$$dG_0/d\tau = -K_0, \qquad dK_0/d\tau = 2K_0(G_0 - 1) \qquad (11.61)$$

This system of equations has a first integral

$$K_0(\tau) = 2G_0(\tau) - G_0^2(\tau) + C_3 \qquad (11.62)$$

where C_3 is an integration constant. It turns out that the knowledge of this invariant is sufficient to solve our problem. Indeed, at the end of the pulse (point P), the pulse has to match the outer solution with $\sigma = 0$: $I_1 = 0$ and $\mathcal{E}_2(0) = B_0(0) = \mathcal{E} + C_1$. At the beginning of the pulse (point Q), the pulse has to match the outer solution with $\sigma = p$: $I_1 = 0$ and $\mathcal{E}_2(p) = B_0(p) = \mathcal{E} + C_1 e^{-p}$. Inserting these relations in the expression for the invariant (11.62) leads to

$$2(\mathcal{E} + C_1) - (\mathcal{E} + C_1)^2 + C_3 = 0$$
$$2(\mathcal{E} + C_1 e^{-p}) - (\mathcal{E} + C_1 e^{-p})^2 + C_3 = 0 \qquad (11.63)$$

These matching conditions determine a closed expression for the two constants

$$C_1 = 2(1 - \mathcal{E})/(1 + e^{-p}), \qquad C_3 = -1 + [(1 - \mathcal{E})\tanh(p/2)]^2 \qquad (11.64)$$

From (11.59) we obtain a first result, namely a closed equation for the pulse period p

$$(\mathcal{E} - 1 - R)p = 2(\mathcal{E} - 1)\tanh(p/2) \qquad (11.65)$$

This equation has solutions only if $\mathcal{E} > 1 + R$. With this expression for the period, we can use the invariant (11.62) and C_3 given by (11.64) to obtain

$$\gamma I_1 \simeq K_0(\tau) = -[G_0(\tau) - 1]^2 + [(1 - \mathcal{E})\tanh(p/2)]^2 \qquad (11.66)$$

Hence the peak intensity is reached for a time $\tau_{\max}$ such that $G_0(\tau_{\max}) = 1$ and is given by

$$\max(I_1) = [(1 - \mathcal{E})\tanh(p/2)]^2/\gamma \tag{11.67}$$

Thus this application of the method of matched asymptotic expansions gives a simple equation (11.65) for the pulse period in terms of which the peak intensity (11.67) is also expressed. In [11] it is shown that the analytic results for p and $\max(I_1)$ derived in this section are in excellent agreement with a numerical simulation of the dynamical equations (11.41). The same agreement was found in [9] for the LSA.

11.4 Relaxation times in nonlinear optics

At this point, it is perhaps adequate that we review the concept of characteristic time.

The basic relaxation times were introduced phenomenologically in chapter 1. They are the photon lifetime $1/\gamma_c$, the population inversion lifetime $1/\gamma_\parallel$ and the atomic polarization or coherence lifetime $1/\gamma_\perp$. In chapter 1, all three times were divided by $\gamma_\perp$ to make them dimensionless. If we express the Maxwell-Bloch equations (1.48)–(1.50) without any scaling (as we did for a LSA at the beginning of this chapter), we obtain

$$dE_0/dt + \gamma_c E_0 = -(N\omega_c\mu/2\varepsilon_0)P_0 \tag{11.68}$$

$$dP_0/dt + [\gamma_\perp + i(\omega_a - \omega_c)]P_0 = -(\mu/\hbar)E_0\mathcal{D} \tag{11.69}$$

$$d\mathcal{D}/dt + \gamma_\parallel(\mathcal{D} - \mathcal{D}_a) = (\mu/2\hbar)(P_0^*E_0 + P_0E_0^*) \tag{11.70}$$

In the absence of light-matter coupling ($\mu = 0$), the three parameters $1/\gamma_c$, $1/\gamma_\perp$, and $1/\gamma_\parallel$ are obviously the relaxation times of the system. The time evolution of the dynamical variables can be written in that case as

$$E_0 \sim \exp(-\gamma_c t), \quad P_0 \sim \exp\{-[\gamma_\perp + i(\omega_a - \omega_c)]t\}, \quad \mathcal{D} \sim \exp(-\gamma_\parallel t) \tag{11.71}$$

The question which arises is: to what extent can we still consider $1/\gamma_c$, $1/\gamma_\perp$, and $1/\gamma_\parallel$ as characteristic times of the interacting system? The relevance of this question lies in the fact that several approximation schemes rely on the implicit assumption that the dynamical variables have a dominant time behavior which is still given by (11.71) even in the presence of interactions. For instance, there is a classification of lasers into class A, B, or C, based on whether one can reduce their description to one, two or three dynamical equations, respectively. In this approach, it is assumed that class A lasers are defined by the good cavity limit conditions $\gamma_c \ll \gamma_\perp$ and $\gamma_c \ll \gamma_\parallel$. On this basis, the atomic variables are adiabatically eliminated and the laser is described by a single equation for the electric filed. Class B lasers fulfill the rate equation approximation $\gamma_\perp \gg \gamma_c$ and $\gamma_\perp \gg \gamma_\parallel$. In this limit, the atomic polarization is adiabatically eliminated and the laser is described by a pair of coupled rate equations for the field and the population inversion. Finally, class C lasers are those for which no adiabatic elimination scheme is possible. These lasers are described by the complete Maxwell-Bloch equations. The relevance of this classification is in the dynamical properties of lasers of different classes. Class A lasers have only steady states in the long time limit. Their response to small external perturbations is of the form $\exp(\lambda t)$ with λ real. Class B lasers may reach steady or periodic states in the long time limit. Their response to small external perturbation is of the form $\exp(\lambda t)$ with λ real or complex. Finally, class C lasers may display practically the whole range of long time solutions, including chaotic states.

In many instances, this A, B, and C classification is useful and the three parameters γ_c, $\gamma_\perp$, and $\gamma_\parallel$ are indeed good guides for the intuition. However, we must refine this analysis because it fails to cover all physically relevant situations. The first problem which is obvious with the above definition of class A, B, and C lasers is that it does not completely specify the magnitude of the three characteristic times. For instance, in the good cavity limit which corresponds to the class A laser, the ratio $\gamma_\perp/\gamma_\parallel$ is still undefined. It can diverge, remain finite or vanish as γ_c becomes much smaller than $\gamma_\perp$ and $\gamma_\parallel$. To analyze this problem, we introduce the variables

$$E_0(t) = (\hbar/\mu)\sqrt{\gamma_\parallel \gamma_\perp}E(t), \quad P_0(t) = \mathcal{D}_a\sqrt{\gamma_\parallel/\gamma_\perp}P(t), \quad \mathcal{D}(t) = \mathcal{D}_a D(t)$$
$$(11.72)$$

and the parameters $A = N\omega_c\mu^2\mathcal{D}_a/(2\hbar\varepsilon_0\gamma_c\gamma_\perp)$ and $\delta_{ac} = (\omega_a - \omega_c)/\gamma_\perp$. In terms of these scaled variables, the Maxwell-Bloch equations (11.68)–(11.70) become

$$dE/dt = -\gamma_c(E + AP) \tag{11.73}$$
$$dP/dt = -\gamma_\perp[(1 + i\delta_{ac})P + ED] \tag{11.74}$$
$$dD/dt = -\gamma_\parallel[D - 1 + (E^*P + EP^*)/2] \tag{11.75}$$

Following the analyses of the good and bad cavity limits in sections 2.2.4, 4.3.1, and 4.3.2 and the rate equation limit in chapter 7, it is easy to verify that the definition of classes A and B requires additional specifications. Lasers of class A have to verify the condition $\varepsilon_1 = \gamma_\perp/\gamma_c \ll 1$ and the requirement that $\gamma_\perp/\gamma_\parallel$, A, δ_{ac}, E, P and D are $\mathcal{O}(1)$ functions with respect to ε_1. In a similar fashion, lasers of class B have to verify the inequality $\varepsilon_2 = \gamma_c/\gamma_\perp \ll 1$ and the requirement that $\gamma_c/\gamma_\parallel$, A, δ_{ac}, E, P and D are $\mathcal{O}(1)$ functions with respect to ε_2.

Although the classification in A, B, and C is far from being exhaustive, we shall not discuss its possible extension but rather recall a number of examples discussed in this book where other time scales arose. In doing so, we shall see that characteristic times can be nonlinear functions of the γ_c, $\gamma_\perp$, and $\gamma_\parallel$, functions of the control parameter A, functions of the initial conditions and even functions of the dynamical variables of the system, i.e., functions of the time!

The first example is the passive Q-switching studied in the previous section. Starting from a problem which has four characteristic times γ_{c1}, γ_{c2}, $\gamma_\perp$, and $\gamma_\parallel$, we have found a domain where the dynamical variables have *two* scalings and where there are *two* different time scales. In terms of the small parameter $\gamma = \gamma_{c2}/\gamma_{c1}$, the intensity is $\mathcal{O}(\gamma)$ between the pulses and evolves on the time scale $\sigma = \gamma\tau = \gamma_{c2}t$. During the pulses, the intensity is $\mathcal{O}(1/\gamma)$ and evolves on the time scale $\tau = \gamma_{c1}t$. The Maxwell-Bloch equations also display passive Q-switching. However, its analysis is more difficult than for the DOPOSA. It was shown analytically in [10], that equations (11.73)–(11.75) on resonance ($\delta_{ac} = 0$) display passive Q-switching in the bad cavity limit $\varepsilon \equiv \gamma_\perp/\gamma_c \ll 1$, with $\gamma_\perp = \gamma_\parallel$ and $A = \mathcal{O}(1/\varepsilon)$. Between pulses, E and P are $\mathcal{O}[\exp(-1/\varepsilon)]$, D is $\mathcal{O}(1)$ and the time scale is $\varepsilon\gamma_\perp t = (\gamma_\perp^2/\gamma_c)t$. During the pulses, $E = \mathcal{O}(\varepsilon^{-1/2})$, $P = \mathcal{O}(\varepsilon^{1/2})$, $D = \mathcal{O}(\varepsilon)$ and the time scale is $\gamma_\perp t$. It is quite clear that passive Q-switching cannot be included in the A, B, and C classification because this form of laser dynamics is dominated by two unstable steady states between which the laser oscillates, regularly or chaotically. The constraint on the parameters, such as $\gamma \ll 1$ for the DOPOSA and $A = \mathcal{O}(1/\varepsilon)$ for the laser, expresses the fact that the two steady states are far away from each other. In that case, the nonlinear nature of the light-matter interaction yields different scalings for the two attractors and therefore two different scalings for the dynamical variables. This in turn induces different time scales for the pulse duration and for the period. The A, B, and C classification is not able to cope with a system displaying a double scaling for the dynamical variables and for the characteristic time.

A second example which is rich in different time scales is the dynamics of OB close to the limit point where the osculating parabola provides a good approximation. This equation was studied in section 4.4. The basic equation is

$$dx/dt = x^2 - 2x + \mu \qquad (11.76)$$

where x is proportional to the cavity field amplitude and μ is proportional to the input field amplitude. The characteristic time in the linearized approximation is $1/2$. However, we have seen in section 4.4.1 that if the control parameter μ is close to its limit point value 1, there is critical slowing down and the characteristic time becomes $1/(2\sqrt{1 - \mu})$. Furthermore, if the initial condition is close to the unstable branch, $x(0) = x_+ \pm \beta$ with $|\beta| \ll 1$ and $1 - \mu = \mathcal{O}(1)$, there is noncritical slowing down and the characteristic time becomes $\ln(|\beta|)/(2\sqrt{1 - \mu})$. In this example, the characteristic time depends essentially on either the control parameter or on the initial condition. In both cases, the characteristic time of the linearized theory has become irrelevant. The value of this example stems from the fact that a exact solution has been obtained. Therefore the full stability analysis has been performed, which determines the long time behavior of the system, given its initial condition and a point in parameter space.

Nonlinear dependence on the characteristic times of the uncoupled system and/or on the control parameter is not new in nonlinear optics. For instance, it is well known that the single mode rate equations of a ring laser (7.52) is characterized by a pair of complex roots (7.53) which yield two characteristic times: oscillations with a period $2\pi/[2\gamma_c\gamma_\parallel(A - 1)]^{1/2}$ which are damped over a decay time $2/(\gamma_\parallel A)$. Here again we find that the decay rates of the uncoupled system are not useful guides.

A final example is drawn from chapter 4 where we have shown that in the limit of fully developed bistability, an asymptotic analysis can still be performed in the vicinity of each limit point. We have shown in section 4.5 that for each limit point there is a reference problem which is a harmonic oscillator with a slow time-dependent frequency. For the lower limit point, the time scale T is given by (4.70). Dividing this time scale by t yields the characteristic decay rate

$$\gamma(t) = (1/t)\int_0^t \sqrt{2D_0(s)/\varepsilon}\,ds \qquad (11.77)$$

This leads to a characteristic time $1/\gamma(t)$, which depends on the time average of the dynamical variable $D_0(t)$ and is, in general, a function of time.

This analysis indicates that nonlinearities may deeply modify the dynamics of a system, to the extent that the effective time scale(s) are nonlinear functions of the time scales of the uncoupled system and of additional parameters and variables.

References

[1]   N. B. Abraham, L. A. Lugiato, and L. M. Narducci, eds, *Instabilities in active optical media, J. Opt. Soc. Am. B* **2** (1985) 1–264.

[2]   B. Zambon, *Phys. Rev. A* **44** (1991) 688.

[3]   G. J. Lasher, *Solid State Electron.* **7** (1964) 707.

[4]   V. N. Lisitsyn and V. P. Chebotaev, *Sov. Phys. JETP* **27** (1968) 227.

[5]   R. W. Hellwarth, In Advances in quantum electronics, J. Singer, ed, p. 334 (Columbia University Press, New York, 1961).

[6]   V. I. Borodulin, N. A. Ermakova, L. A. Rivlin, and V. S. Shil'dyaev, *Sov. Phys. JETP* **21** (1965) 563. According to an author of this paper, the experimental work was performed in 1963 but declassified in late 1964 only. Furthermore, a patent was applied for by L. A. Rivlin (USSR Patent # 166149) on July 3, 1963, on the LSA as a source of pulsed light.

[7]   T. Erneux and P. Mandel, *Z. Physik B* **44** (1981) 353 and 365.
[8]   N. B. Abraham, P. Mandel, and L. M. Narducci, Dynamical instabilities and pulsations in lasers in *Progress in Optics,* Vol. XXV, pp. 1–190, E. Wolf, ed. (North-Holland, Amsterdam, 1988).
[9]   T. Erneux, *J. Opt. Soc. Am. B* **5** (1988) 1063.
[10]   A. C. Fowler and M. J. McGuinness, *Physica* **5D** (1982) 149.
[11]   R.-D. Li, P. Mandel, and T. Erneux, *J. Opt. Soc. Am. B* **8** (1991) 1835.

12

Transverse effects in optical bistability

In this chapter, we remove one assumption that has been implicit since the beginning of this book, namely the unidimensional aspect of the cavity. We now take into account the transverse variation of the field in the resonant cavity. In dealing with transverse effects, there are two possible approaches, very much like in the multimode optical bistability (OB) studied in Chapter 6. One possibility is to project the Maxwell–Bloch equations on a suitable basis. In the introduction of Chapter 6, the difficulty of selecting this suitable basis was explained for 1-D cavities. This difficulty is amplified by the transverse dimensions, especially because of the lateral boundaries. The other approach is to derive global equations (that are still partial differential equations) for slowly varying amplitudes. They are generally variants of well-known nonlinear partial differential equations of mathematical physics, and therefore a large number of results are directly available. However, when transferring results from another domain to optics, some care must be exercised because the relevant domains of parameters are not always compatible. The classic example is the canonical set of parameters for the Lorenz equations that includes $b = \gamma_{\parallel}/\gamma_{\perp} = 8/3$ whereas for atomic transitions the upper bound of b is 2. The reader will find a wealth of results, mostly for the modal expansions, in the special issues and reviews [1]–[4]. A review more specifically oriented toward the global amplitude equations is found in [5] and a mathematical study of these equations is presented in [6].

Which of the two approaches, the modal expansion or the global amplitude equations, should be used depends on the nature of the transverse effects. In many experiments, it is clearly seen that the transverse profile involves modes that have the general structure of the empty cavity modes. What is observed is a dynamics ruled by cavity mode–mode interactions. It is obvious that in this case the modal expansion is the natural way to analyze the system. However, other types of transverse structures may occur. The simplest of them are

160

(1) periodic patterns, such as stripes and hexagons, and (2) localized structures corresponding to isolated maxima of the transverse intensity without a relation with any mode structure. In these cases, it is the second method (global amplitude equations) that is more adequate. Let a Fresnel number be defined by $\mathcal{F}_\perp = A_\perp / \lambda_c L$, where $A_\perp$ is the transverse area, λ_c the optical wavelength of the electric field, and L the cavity length. The modal decomposition is better adapted if the Fresnel number is small, implying modes with very different losses, if the cavity mode frequencies are well separated and if the boundary conditions play a dominant role (mirrors with a large curvature). The opposite situation, best described by global amplitude equations, is characterized by a large Fresnel number, a high degeneracy of the mode frequencies, and quasi-planar mirrors.

The derivation of amplitude equations from the Maxwell–Bloch equations is still the subject of debate and criticism, except for one case that is exempt from these unresolved difficulties: This case is nascent OB and is studied in this chapter.

12.1 The adiabatic limit

Our starting point is with equations (4.1)–(4.3)

$$\partial E/\partial t = \gamma_c[-(1 + i\Delta)E - 2CP + E_i + ia\mathcal{L}_\perp E] \qquad (12.1)$$

$$\partial P/\partial t = \gamma_\perp[-(1 + i\delta)P + ED] \qquad (12.2)$$

$$\partial D/\partial t = \gamma_\parallel[1 - D - (1/2)(E^*P + EP^*)] \qquad (12.3)$$

with $a = c/2\gamma_c k_c \simeq L\lambda_c/4\pi T_f$. We have introduced two modifications needed for a clear exposition of the problem:

1. The time t is the physical time so that γ_c, $\gamma_\perp$, and $\gamma_\parallel$ are the unscaled decay rates or inverse decay times.

2. In the equation for the field, we have added the transverse Laplace operator $\mathcal{L}_\perp = \partial^2/\partial x^2 + \partial^2/\partial y^2$.

If we divide the transverse coordinates by the corresponding transverse cavity length, $\overline{x} = x/L_x$ and $\overline{y} = y/L_y$, the transverse Laplace operator becomes

$$a\mathcal{L}_\perp = a_x \, \partial^2/\partial\overline{x}^2 + a_y \partial^2/\partial\overline{y}^2$$

$$a_x \simeq L\lambda_c/4\pi T_f L_x^2 = 1/4\pi T_f \mathcal{F}_x, \qquad a_y \simeq L\lambda_c/4\pi T_f L_y^2 = 1/2\pi T_f \mathcal{F}_y$$

$$(12.4)$$

in terms of the Fresnel numbers $\mathcal{F}_x = L_x^2/\lambda_c L$ and $\mathcal{F}_y = L_y^2/\lambda_c L$. Diffraction effects become major features when these Fresnel numbers are larger than unity. To determine which is the difficulty associated with the transverse effects, let us do some "fast physics" and perform a naive adiabatic elimination of the atomic variables, assuming that the good cavity conditions are satisfied: $\gamma_c \ll \gamma_\parallel$ and $\gamma_c \ll \gamma_\perp$. In this case, we expect to be justified in neglecting the left-hand side of the material equations (12.2) and (12.3) to obtain a pair of algebraic equations for the atomic polarization and population difference: $P = ED/(1 + i\delta)$ and $D = 1 - (1/2)(E^*P + EP^*)$. Solving these equations for P and D and inserting the result in the field equation (12.1) gives

$$\partial E/\partial t = \gamma_c \left[-(1 + i\Delta)E - \frac{2C(1 - i\delta)E}{1 + \delta^2 + |E|^2} + E_i + ia\mathcal{L}_\perp E \right]. \quad (12.5)$$

As it stands, this equation is well defined. It is essentially a diffraction equation as might be expected from Maxwell's equation for the electric field. The trouble is precisely with the diffraction term. As shown in Section 4.3.1, equation (12.5) is the dominant order contribution of an asymptotic expansion in the small parameter γ_c/γ (whichever of the two atomic decay rates is chosen for γ is irrelevant for this discussion). In the spirit of this approach, the next order contribution will be proportional to γ_c/γ and includes $\partial P/\partial t$ and $\partial D/\partial t$. These terms will contribute to the transverse Laplace operator, but lead to a complex coefficient that becomes $a_1 + ia_2$. This has the effect of transforming the diffraction equation into a diffusion–diffraction equation and the sign of the diffusion coefficient determines the stability of the equation. In this way the stability of the solutions is determined by a coefficient that does not appear in the dominant order equation (12.5). This is a case of singular perturbation where the highest order derivative that determines the stability properties is multiplied by the smallness parameter. Hence we have reasons to expect that an expansion in powers of the small parameter γ_c/γ is not regular and that a more sophisticated treatment is necessary. This difficulty has recently been recognized but not cured yet in the general case.

12.2 Nascent optical bistability

The difficulty we are facing is that all we have at present is negative information: An expansion of E, P, and D in powers of γ_c/γ is unlikely to converge. However, we have seen at many places in this book that new expansions emerge when there is critical slowing down. These expansions do not rely on the existence of inequalities among the atomic and cavity decay rates leading to an obvious classification into slow and fast variables. The clearest example was in Chapter 4, where we studied the limit $C \to \infty$ (fully developed hysteresis). Despite the fact that the three decay rates γ_c, $\gamma_\perp$, and $\gamma_\parallel$ were of the same order of magnitude, a single evolution equation was derived in the vicinity of each limit point. Another candidate for this type of analysis is found in Section 5.3.1, where we proved the occurrence of critical slowing down around the point of inflexion with vertical slope $(n_c, I_c, \theta_c) = (2/\sqrt{3}, (2/\sqrt{3})^3, \sqrt{3})$ in OB. Although this property was derived in the bad cavity limit, the conclusion about the existence of critical slowing down can be related to geometrical considerations only and therefore is not affected by the bad cavity limit. In this section, we consider the limit of an inflexion point with vertical slope and comparable decay rates. This is the nascent hysteresis limit of OB. In this case we expect the derivation of the field equation (12.5) to be simply wrong.

The homogeneous steady state solution $E = E(E_i)$ of (12.1)–(12.3) satisfies the implicit equation

$$E_i = E\left(1 + i\Delta + \frac{2C(1 - i\delta)}{1 + \delta^2 + |E|^2}\right) \tag{12.6}$$

It has an inflexion point whose coordinates are

$$|E_{\text{infl}}|^2 = -(1 + \delta^2) + \frac{4}{9}\frac{(C - 1 + \delta\Delta)^2}{1 + \Delta^2} \tag{12.7}$$

$$E_{i,\text{infl}}^2 = -(1 + \delta^2)(1 + \Delta^2) + 4C(1 - \delta\Delta) + \frac{4}{3}(C - 1 + \delta\Delta)^2 \tag{12.8}$$

which are determined by the condition $\partial E/\partial E_i = \infty$ along the steady state solution (12.6). Bistability appears if the parameter C satisfies the relation

$$27C_c(1 + \Delta^2)(1 + \delta^2) = 4(C_c - 1 + \Delta\delta)^3 \tag{12.9}$$

that originates from the double condition $\partial E/\partial E_i = \infty$ and $\partial^2 E/\partial E_i^2 = \infty$ along the steady state: The two limit points coincide with the inflexion point. If we impose the restriction $\Delta = -\delta$, equation (12.9) simplifies and has one real positive solution

$$C_c = 4(1 + \delta^2) \tag{12.10}$$

The condition of antisymmetric detuning $\Delta = -\delta$ is equivalent to requiring that the external field frequency be that of a laser using the same nonlinear medium and the same cavity: $\omega_i = (\gamma_\perp \omega_c + \gamma_c \omega_a)/(\gamma_c + \gamma_\perp)$. If $C = C_c$, the steady state solution at the inflexion point (12.7)–(12.8) is

$$E_c = (1 + i\delta)\sqrt{3}, \qquad P_c = \sqrt{3}/4, \qquad D_c = 1/4 \tag{12.11}$$

with $E_{i,\mathrm{infl}} \equiv E_{i,c} = 3(1 + \delta^2)\sqrt{3}$. Let us perform a linear stability analysis of this solution. We seek solutions of the form $X(\mathbf{r}, t) = X_c + \eta x(\mathbf{r}, t)$ with $|\eta| \ll 1$, where X is any of the variables E, P, and F, X_c is the corresponding steady state given in (12.11). The variable $\mathbf{r}$ refers to the transverse coordinates. This implies that the parameters C_c and $E_{i,c}$ are frozen. Thus we obtain a set of linear equations for the deviations $x(\mathbf{r}, t)$. To proceed further in this calculation, we need to specify the action of the transverse Laplace operator. Let us assume that the boundary conditions admit the $\exp(i\mathbf{k}\mathbf{r})$ as solutions and that $(\mathcal{L}_\perp + k^2)\exp(i\mathbf{k}\mathbf{r}) = 0$. The linear equations for the $x(\mathbf{r}, t)$ lead to a fifth order characteristic equation

$$P(5, \lambda) = \sum_{n=1}^{5} a_n \lambda^n = 0 \tag{12.12}$$

$a_5 = 1$

$a_4 = -(\gamma_\| + 2\gamma_\perp + 2\gamma_c)$

$a_3 = -(5 + 3\delta^2)\gamma_\|\gamma_\perp - (1 + \delta^2)\gamma_\perp^2 - 2\gamma_\|\gamma_c - 4(2 + \delta^2)\gamma_\perp\gamma_c$
$\quad\quad - (1 + \delta^2 - 2ak^2\delta - a^2k^4)\gamma_c^2$

$a_2 = -(1 + \delta^2 - 2ak^2\delta - a^2k^4)\gamma_\|\gamma_c^2 + 6(1 + \delta^2)\gamma_\perp^2\gamma_c$
$\quad\quad + 4(1 + \delta^2)\gamma_\|\gamma_\perp^2 - [6(1 + \delta^2) - 4ak^2\delta + 2a^2k^4]\gamma_\perp\gamma_c^2$
$\quad\quad + 4(2 + \delta^2)\gamma_\|\gamma_\perp\gamma_c$

$a_1 = (1 + \delta^2)[-9(1 + \delta^2) + 6ak^2\delta - a^2k^4]\gamma_\perp^2\gamma_c^2$
$\quad\quad - 12(1 + \delta^2)\gamma_\|\gamma_\perp^2\gamma_c + [-3(1 + 2\delta^2)^2 + 4ak^2\delta(4 + 3\delta^2)$
$\quad\quad - a^2k^4(5 + 3\delta^2)]\gamma_\|\gamma_\perp^2\gamma_c$

$a_0 = (1 + \delta^2)4ak^2(3\delta - ak^2)\gamma_\|\gamma_\perp^2\gamma_c^2$

The condition $\lambda = 0$ defines the unstable modes at C_c and $E_{i,c}$

$$a^2k_{\mathrm{crit}}^4 = 3\delta ak_{\mathrm{crit}}^2 \tag{12.13}$$

If $\delta = (\omega_a - \omega_i)/\gamma_\perp \leq 0$, the only possible solution is the homogeneous mode $ak_{\mathrm{crit}}^2 = 0$. Otherwise the solutions are $ak_{\mathrm{crit}}^2 = 0$ and $ak_{\mathrm{crit}}^2 = 3\delta$. This is a first indication that transverse effects differ for positive and negative values of δ. In the following sections, we will study the vicinity of the steady solution (12.11). That is, we seek solutions of the equations (12.1)–(12.3) in the form

$$X = X_c + \varepsilon X_1 + \mathcal{O}(\varepsilon^2), \qquad E_i = E_{i,c} + \varepsilon E_{i,1} + \mathcal{O}(\varepsilon^2) \tag{12.14}$$

for $X = E, P,$ or D. The small parameter ε is defined by

$$C \equiv 4(1 + \delta^2) + \alpha \varepsilon^2, \qquad \alpha = \pm 1 \qquad (12.15)$$

Note that (12.14) is not the usual perturbation expansion around a given steady state: The corrections $\mathcal{O}(\varepsilon^n)$ contain a constant part that are corrections to the steady state (12.11). This is because we analyze the deviation from a particular point of the steady state curve and not the deviation from the steady state as a whole. As already explained in preceding chapters, the motivation of this approach is to exploit the critical slowing down that characterizes the vicinity of $(C_c, E_{i,c})$.

12.3 Negative detuning: The Ginzburg–Landau equation

If $\delta < 0$, we expect a rather simple situation because the first mode to become unstable is the homogeneous mode $k_{\mathrm{crit}} = 0$. However, we still have to find space and time scales to proceed with the perturbation expansion. To find this scaling, let us consider a linear stability analysis using (12.14) with C given by (12.15). It is still a stability analysis around the inflexion point, but without frozen parameters. This leads to a fifth-order polynomial $P(5, \lambda, \varepsilon)$ which has the simple expression

$$
\begin{aligned}
P(5, \lambda, \varepsilon) &= \sum_{n=1}^{5} [a_n + \alpha\varepsilon^2 b_n + \alpha^2\varepsilon^4 c_n]\lambda^n \\
&= P(5, \lambda) - \alpha\varepsilon^2\lambda^3\gamma_\perp\gamma_c - \alpha\varepsilon^2\lambda^2\gamma_\perp\gamma_c(\gamma_c + \gamma_\perp + \gamma_\parallel/2) \\
&\quad + \alpha\varepsilon^2\lambda[(-7 - 9\delta^2 + ak^2\delta)\gamma_\perp\gamma_c + (1 - 3\delta^2 + 3ak^2\delta)\gamma_\parallel\gamma_\perp\gamma_c^2/2 \\
&\quad - \gamma_\parallel\gamma_\perp^2] + \alpha^2\varepsilon^4\lambda\gamma_\perp^2\gamma_c^2/4 + \alpha\varepsilon^2[6(1 + \delta^2) + ak^2\delta \\
&\quad + \alpha\varepsilon^2]\gamma_\parallel\gamma_\perp^2\gamma_c^2/2
\end{aligned}
\qquad (12.16)
$$

where $P(5, \lambda)$ is the polynomial defined by (12.12). Let us fix $\delta < 0$ with $|\delta| = \mathcal{O}(1)$. For $\varepsilon = 0$, λ vanishes if $ak_{\mathrm{crit}}^2(ak_{\mathrm{crit}}^2 - 3\delta) = 0$. In the limit $\varepsilon \to 0$, equation (12.16) can be written as

$$\lambda P(4, \lambda, \varepsilon) + (1 + \delta^2)4ak^2(3\delta - ak^2) + \alpha\varepsilon^2[3(1 + \delta^2) + ak^2\delta/2] = 0$$

$$(12.17)$$

where $P(4, \lambda, \varepsilon)$ is a quartic in λ. No coefficient of the quartic vanishes as $\varepsilon = 0$. This equation indicates that at least one root scales like $\lambda = \mathcal{O}(\varepsilon^2)$ if $k = \mathcal{O}(\varepsilon)$. With this scaling, the space and time scales contribute to the same order and may balance each other. Therefore, we seek corrections to the steady state at criticality that depend on time and space through the slow variables

$$\tau = \varepsilon^2 t, \qquad \mathbf{v} = \varepsilon \mathbf{r} \qquad\qquad (12.18)$$

or equivalently $\mathcal{L}_\perp = \varepsilon^2 \mathcal{L}_2$. Thus, if X is any of the functions E, P, and D, we seek solutions of the form

$$X(t, \mathbf{r}, \varepsilon) = X_c + \varepsilon X(\tau, \mathbf{v}, \varepsilon) = X_c + \varepsilon X_1(\tau, \mathbf{v}) + \mathcal{O}(\varepsilon^2)$$

$$E_i(\varepsilon) = E_{i,c} + \varepsilon E_{i,1} + \mathcal{O}(\varepsilon^2) \qquad\qquad (12.19)$$

and $\partial X/\partial t = \varepsilon^2 \partial X/\partial \tau$. The first-order equations are

$$0 = E_{i,1} - (1 - i\delta)E_1 - 8(1 + \delta^2)P_1,$$

$$0 = -(1 + i\delta)P_1 + \sqrt{3}(1 + i\delta)D_1 + E_1/4,$$

$$0 = D_1 + \sqrt{3}(E_1 + E_1^*)/8 + \sqrt{3}[(1 + i\delta)P_1^* + (1 - i\delta)P_1]/2$$

$$(12.20)$$

The solution of this set of algebraic equations is

$$E_1 = B_1/(1 - i\delta), \qquad P_1 = -B_1/[8(1 + \delta^2)]$$

$$D_1 = -\sqrt{3}B_1/[8(1 + \delta^2)], \qquad E_{i,1} = 0 \qquad\qquad (12.21)$$

The only constraint on the function B_1 is that it must be real. It is this reality condition that will eventually lead to an amplitude equation for B_1. The $\mathcal{O}(\varepsilon^2)$ equations lead also to a simple algebraic problem whose solution is

$$E_2 = B_2/(1 - i\delta), \qquad P_2 = -B_2/[8(1 + \delta^2)]$$

$$D_2 = -\sqrt{3}B_2/[8(1 + \delta^2)] + B_1^2/[8(1 + \delta^2)^2] \qquad E_{i,2} = \alpha\sqrt{3}/2$$

$$(12.22)$$

Again the only constraint on B_2 is that it must be a real function. Thus we have to examine the next order problem, which is

$$\partial E_1/\partial \tau = \gamma_c[E_{i,3} - (1 - i\delta)E_3 - 8(1 + \delta^2)P_3 - 2\alpha P_3 + ia\mathcal{L}_2 E_1]$$

$$\partial P_1/\partial \tau = \gamma_\perp[-(1 + i\delta)P_3 + \sqrt{3}(1 + i\delta)D_3 + E_3/4 + E_1 D_2 + E_2 D_1]$$

$$\partial D_1/\partial \tau = \gamma_\parallel\{-D_3 - \sqrt{3}(E_3 + E_3^*)/8 - \sqrt{3}[(1 + i\delta)P_3^* + (1 - i\delta)P_3]/2$$

$$- (P_1^* E_2 + P_2^* E_1 + P_1 E_2^* + P_2 E_1^*)/2\} \qquad\qquad (12.23)$$

Eliminating P_3 and D_3 from the last two equations, the first equation leads to an algebraic equation for E_3. From the imaginary part of this algebraic equation we obtain

$$E_3 = \frac{B_3}{1 + i\delta} + \frac{i(1 + i\delta)}{3(1 + \delta^2)^2}\left[\delta(\frac{1}{\gamma_\perp} - \frac{1}{\gamma_c})\partial B_1/\partial \tau + a\mathcal{L}_2 B_1\right]$$

$$(12.24)$$

with B_3 real, whereas the real part is

$$\frac{\beta(\delta)}{4(1 + \delta^2)}\partial B_1/\partial \tau = E_{i,3} + \frac{\alpha B_1}{4(1 + \delta^2)} - \frac{B_1^3}{4(1 + \delta^2)^2} - \frac{a\delta}{1 + \delta^2}\mathcal{L}_2 B_1$$

(12.25)

with the definition

$$\beta(\delta) = \frac{4}{\gamma_c} + \frac{3(1 + \delta^2)}{\gamma_\parallel} + \frac{1 + 3\delta^2}{\gamma_\perp}$$

(12.26)

The main result of this section is the Ginzburg–Landau equation (12.25). We notice that to dominant order, the equation that rules the dynamics of the system is of the diffusive type and not of the diffractive type. Diffraction appears in the third-order contribution E_3 as shown in the solution (12.24), whereas diffusion dominates the first-order term E_1. Equation (12.25) may have bounded solutions only if the diffusion coefficient $-a\delta/(1 + \delta^2)$ is positive. This severely restricts the domain of validity of the Ginzburg–Landau equation and introduces the next question: Which equation has to be used if δ vanishes or is positive? This question is considered in the next sections.

To close this section, let us notice that we can write the Ginzburg–Landau equation in terms of the deviation $\mathcal{E} = E - E_c$ from the steady electric field amplitude

$$\beta\ \partial\mathcal{E}/\partial t = 4(1 + i\delta)\mathcal{Y} + \mathcal{E}(C - |\mathcal{E}|^2) - 4a\delta\mathcal{L}_\perp\mathcal{E}$$

(12.27)

with the parameters

$$C = 4(1 + \delta^2) + \mathcal{C},\qquad E_i = 3(1 + \delta^2)\sqrt{3} + \frac{\sqrt{3}}{2}\mathcal{C} + \mathcal{Y}$$

(12.28)

The constraints $|\mathcal{E}| \ll 1, \mathcal{C} \ll 1$, and $\mathcal{Y} \ll 1$ still apply to (12.27) and it would be a gross mistake to use the Ginzburg–Landau equation (11.26) beyond that domain of parameters. From the steady solution (12.11), it is clear that the complex term $1 + i\delta$ in the Ginzburg–Landau equation (12.27) is a signature of the phase difference between the input field and the cavity field because $E_{i,c} = 3(1 - i\delta)E_c$.

12.4 Small detuning: The Swift–Hohenberg equation

In the previous section, we have derived a complex Ginzburg–Landau equation that makes sense only if $\delta < 0$. In the absorptive limit $\delta = 0$, the diffusion coefficient vanishes. Therefore, we have at least four more domains to analyze: the purely absorptive case ($\delta = 0$), the domain of positive detuning ($\delta > 0$)

and the two boundary layers, where δ is small and either positive or negative. It turns out that the two boundary layers and the absorptive limit can be described by a unique equation. To find the space and time scalings in the boundary layers, let us consider the domain $|\delta| = \mathcal{O}(\varepsilon)$. From (12.17) a balance between the space and time scales is realized for $\lambda = \mathcal{O}(\varepsilon^2)$ and $k^2 = \mathcal{O}(\varepsilon)$. This suggests that we consider the scaling

$$\tau = \varepsilon^2 t, \qquad \mathbf{v} = \varepsilon^{1/2}\mathbf{r}, \qquad \delta = \varepsilon\delta_1 \qquad (12.29)$$

and therefore $\mathcal{L}_\perp = \varepsilon\mathcal{L}_1$. This scaling shifts the most unstable wave numbers toward the origin so that a series expansion in powers of ε around the homogeneous steady state solution (12.11) still captures both the homogeneous and the inhomogeneous unstable modes. We keep the expansions (12.19) though with the new definition of the space scale.

To first order in ε, we obtain the equations

$$0 = E_{i,1} - E_1 - 8P_1$$
$$0 = -P_1 + \sqrt{3}D_1 + E_1/4$$
$$0 = D_1 + \sqrt{3}(E_1 + E_1^*)/8 + \sqrt{3}(P_1^* + P_1)/2 \qquad (12.30)$$

whose solution is

$$E_1 = B_1, \qquad P_1 = -B_1/8, \qquad D_1 = -\sqrt{3}B_1/8, \qquad E_{i,1} = 0 \quad (12.31)$$

with B_1 real. The second-order equations have the solution

$$E_2 = B_2 + i\delta_1 E_1 + ia\mathcal{L}_1 E_1/3, \qquad P_2 = -B_2/8 + ia\mathcal{L}_1 E_1/12$$
$$D_2 = (-\sqrt{3}B_2 + E_1^2)/8, \qquad E_{i,2} = \alpha\sqrt{3}/2 \qquad (12.32)$$

where B_2 has to be real. Note that diffraction already appears in this second-order expression. Finally, the third-order equations are

$$\partial E_1/\partial\tau = \gamma_c[E_{i,3} - E_3 - 8P_3 - 2\alpha P_1 + i\delta E_2 - 8\delta_1^2 P_1 + ia\mathcal{L}_1 E_2]$$
$$\partial P_1/\partial\tau = \gamma_\perp[-P_3 + \sqrt{3}(D_3 + i\delta_1 D_2) + E_3/4 + E_1 D_2 + E_2 D_1 - i\delta_1 P_2]$$
$$\partial D_1/\partial\tau = \gamma_\parallel\{-D_3 - \sqrt{3}(E_3 + E_3^*)/8 - \sqrt{3}[P_3^* + P_3 + i\delta_1(P_2^* - P_2)]/2$$
$$- (P_1^* E_2 + P_2^* E_1 + P_1 E_2^* + P_2 E_1^*)/2\} \qquad (12.33)$$

Solving the last two equations for P_3 and D_3, we obtain from the first equation a closed equation for E_3. The imaginary part of this equation leads to a condition for E_3 that we do not need here, and the real part leads to an equation for the real function $B_1 = E_1 = \mathcal{E}/\varepsilon$

$$\beta_0\,\partial\mathcal{E}/\partial t = 4\mathcal{Y} + \mathcal{E}(C - \mathcal{E}^2) - 4a\delta\mathcal{L}_\perp\mathcal{E} - \frac{4}{3}a^2\mathcal{L}_\perp\mathcal{L}_\perp\mathcal{E} \qquad (12.34)$$

The parameters $\mathcal{Y}$ and $\mathcal{C}$ are defined by (12.28) and $\beta_0 = \beta(0)$ is defined by (12.26). This equation holds for small δ, positive or negative, and remains valid if $\delta = 0$. Equation (12.34) describes a combination of three effects that may compete with each other:

- The algebraic term contains a linear and a cubic term. If $\mathcal{C} > 0$, the linear term induces a divergence that is balanced by the cubic term.
- The Laplace operator is multiplied by a "diffusion" coefficient that is negative if $\delta > 0$. In that case, this term contributes to the divergence of the solutions.
- The square of the Laplace operator that is multiplied by a constant negative coefficient compensates the divergence induced by the linear Laplace operator.

Hence for both the algebraic contributions and the transverse operators, the linear term may have a divergence that is compensated by the nonlinear term. Let us emphasize that the Swift–Hohenberg equation (12.34) is valid only in the domain $|\mathcal{E}| \ll 1$, $\mathcal{C} \ll 1$, $\mathcal{Y} \ll 1$, and $\delta^2 \ll 1$. Its extension outside of this domain is not justified. For analytical studies, it is important to remark that the Swift–Hohenberg equation derived in this section has real coefficients, whereas the Ginzburg–Landau equation derived for negative detuning has complex coefficients. Real coefficients make the analytic study of the Swift–Hohenberg equation much simpler.

It was pointed out that the derivation presented in this section indicates a relation between the Ginzburg–Landau and the Swift–Hohenberg equations [7]. Indeed, we can apply the derivation followed in this section after performing an adiabatic elimination of the atomic variables. Assuming $\gamma_c \ll \gamma_\parallel$ and $\gamma_c \ll \gamma_\perp$, the basic equations (12.1)–(12.3) yield (12.5), which is a Ginzburg–Landau equation with a saturable nonlinearity and a diffraction transverse operator. Applying to this equation the perturbation expansion used in this section leads to the Swift–Hohenberg equation (12.34) with $\beta_0 = 4/\gamma_c$. Hence the diffractive Ginzburg–Landau equation with a complex saturable nonlinearity leads, close to the instability threshold, to a diffusive Swift–Hohenberg equation with a real nonlinearity. This indicates that the properties of the Ginzburg–Landau equation may vary significantly with the type of nonlinearity. It also suggests that the opposition diffusion versus diffraction is not relevant in this case.

If the detuning δ does not vanish, we can introduce the change of variables

$$\mathcal{E} = \mathcal{E}_0 + u\delta\sqrt{3}, \qquad 4\mathcal{Y} + (\alpha\varepsilon^2 - \mathcal{E}_0)\mathcal{E}_0 = 0,$$
$$\tau = 3\delta^2 t/\beta_0, \qquad \mathbf{r}' = 2a\mathbf{r}/3\delta \tag{12.35}$$

In the new variables, equation (12.34) takes the classic form

$$\partial u/\partial \tau = pu + qu^2 - u^3 - (1 + \mathcal{L}_\perp)^2 u \qquad (12.36)$$

where $p = 1 + (\mathcal{C} - 3\mathcal{E}_0^2)/3\delta^2$ and $q = -\mathcal{E}_0\sqrt{3}/\delta$. The traditional Swift–Hohenberg equation has $q = 0$. However, a number of general features are not modified by the fact that $q \neq 0$, though it destroys the invariance of the equation under the transformation $u \to -u$. In Section 12.6 we will consider some properties of equation (12.36).

12.5 Analysis of the Turing bifurcation

In this section, we analyze the spatially inhomogeneous solutions of the Swift–Hohenberg equation (12.34) that we write as

$$\partial X/\partial t = 4\mathcal{Y} + X(\mathcal{C} - X^2) - 4\delta\mathcal{L}_\perp X - \frac{4}{3}\mathcal{L}_\perp\mathcal{L}_\perp X \qquad (12.37)$$

by rescaling the space and time variables. The homogeneous steady state satisfies the equation $4\mathcal{Y} + X(\mathcal{C} - X^2) = 0$, which has two limit points, where $\partial \mathcal{Y}/\partial X = 0$

$$X_{L\pm} = \pm\sqrt{\mathcal{C}/3}, \qquad \mathcal{Y}_{L\pm} = \mp(\mathcal{C}/6)\sqrt{\mathcal{C}/3} \qquad (12.38)$$

Thus there is a domain of bistability if $\mathcal{C} > 0$. Otherwise the steady solution is monostable.

12.5.1 Linear stability analysis

The linear analysis proceeds as usual. We seek solutions of the Swift–Hohenberg equation (12.37) that are of the form $X(\mathbf{r}, t) = X_s + \eta u(\mathbf{r}, t) + \mathcal{O}(\eta^2)$, where X_s is the real solution of $4\mathcal{Y} + X(\mathcal{C} - X^2) = 0$ and $0 < \eta \ll 1$. To first order in η, we obtain

$$\partial u/\partial t = (\mathcal{C} - 3X_s^2)u - 4\delta\mathcal{L}_\perp u - \frac{4}{3}\mathcal{L}_\perp\mathcal{L}_\perp u \qquad (12.39)$$

This equation admits solutions of the form $u(\mathbf{r}, t) \propto \exp(i\mathbf{kr} + \lambda t)$ because we have already assumed that the Laplace operator has the property $(\mathcal{L}_\perp + k^2)\exp(i\mathbf{kr}) = 0$. This leads to the characteristic equation

$$\lambda = \mathcal{C} - 3X_s^2 + 4\delta k^2 - 4k^4/3 \qquad (12.40)$$

The opposite roles of the linear and cubic terms, and of the Laplace and squared Laplace operators, are manifest through the sign of their respective contributions to the characteristic equation. The vanishing of λ defines a band of unstable wave numbers bounded by

$$k_\pm^2 = \frac{3}{2}\left(\delta \pm \sqrt{\delta^2 - X_s^2 + C/3}\right) \qquad (12.41)$$

This band of unstable modes reduces to a single point when the two bounds of the domain coincide. This is the so-called Turing instability at which the homogeneous steady solution becomes unstable and a inhomogeneous steady solution emerges. At the Turing instability, we therefore have

$$X_{T\pm} = \pm\sqrt{\delta^2 + C/3}, \qquad \mathcal{Y}_{T\pm} = \pm\frac{3\delta^2 - 2C}{12}\sqrt{\delta^2 + C/3}, \qquad k_T^2 = 3\delta/2$$
$$(12.42)$$

This instability can occur in the homogeneous monostable and bistable domains. However, it requires a positive detuning.

Note that in this section we have performed the usual stability analysis around the whole solution of the steady state equation $4\mathcal{Y} + X(C - X^2) = 0$. This is in contrast with the stability analyses around a critical point of the steady state solution that lead to the characteristic equations (12.12) and (12.16).

12.5.2 Nonlinear stability analysis

The problem with the linearized stability analysis in the preceding section is that it gives no indication on the stability of the emerging inhomogeneous solutions. The situation here is quite similar, conceptually, to the situation met in Chapter 10 with the Hopf bifurcation: The linear analysis determines the nature of the emerging solution (time-periodic for the Hopf bifurcation, space-periodic for the Turing instability) but a nonlinear analysis is required to determine the stability of the emerging solution. In fact, the procedure that we follow now parallels the analysis of the Hopf bifurcation in the time domain. We seek solutions of the Swift–Hohenberg equation and of the steady state equation

$$\partial X/\partial t = 4\mathcal{Y} + X(C - X^2) - 4\delta\mathcal{L}_\perp X - \frac{4}{3}\mathcal{L}_\perp\mathcal{L}_\perp X \qquad (12.43)$$

$$4\mathcal{Y} = X_s(X_s^2 - C) \qquad (12.44)$$

in the vicinity of the Turing instability. Therefore the following expansions are introduced

$$X(\mathbf{r}, t) = X_s(\eta) + u(\mathbf{r}, t, \eta)$$
$$X_s(\eta) = X_{T\pm} + \eta a_1 + \mathcal{O}(\eta^2)$$
$$u(\mathbf{r}, t, \eta) = u(\mathbf{r}, \sigma, \eta) = \eta[u_0(\mathbf{r}, \sigma) + \eta u_1(\mathbf{r}, \sigma) + \mathcal{O}(\eta^2)]$$
$$\mathcal{Y} = \mathcal{Y}_{T\pm} + \eta\mathcal{Y}_1 + \mathcal{O}(\eta^2)$$
$$\sigma = \eta^2 t \qquad (12.45)$$

The parameter η is a measure of the deviation from the Turing bifurcation point. The slow time σ signals the critical slowing down that characterizes the Turing bifurcation.

To order η^0, we obtain from (12.43)

$$\mathcal{L}u_0 \equiv (C - 3X_{T\pm}^2 - 4\delta\mathcal{L}_\perp - \frac{4}{3}\mathcal{L}_\perp\mathcal{L}_\perp)u_0 = 0 \qquad (12.46)$$

This equation has solutions of the form

$$u_0 = W(\sigma)\exp(i\mathbf{k}_T\mathbf{r}) + c.c. \qquad (12.47)$$

where the wave vector $\mathbf{k}_T$ is restricted by the condition $|\mathbf{k}_T|^2 = 3\delta/2$ and W is an arbitrary function of the slow time σ.

At order η, the Swift–Hohenberg equation (12.43) yields

$$\mathcal{L}u_1 = 6X_{T\pm}a_1u_0 + 3X_{T\pm}u_0^2 \qquad (12.48)$$

whereas the steady state equation (12.44) yields

$$4\mathcal{Y}_1 = 3\delta^2 Ca_1 \qquad (12.49)$$

The problem that arises with equation (12.48) in space is the same as the problem we have encountered in Section 10.4.1 in the time domain. The homogeneous part of (12.48) has spatial oscillations at the wave vector $\mathbf{k}_T$ and so does the inhomogeneous term. Hence (12.48) has only trivial periodic solutions unless a solvability condition is satisfied: The right-hand side of (12.48) must be orthogonal to the eigenfunctions of $\mathcal{L}^+$, the adjoint of $\mathcal{L}$. Since $\mathcal{L}$ is self-adjoint, u_0 is also the eigenfunction of the adjoint operator. Hence the solvability condition becomes

$$\int_0^{\vec{\lambda}_T} u_0(2a_1u_0 + u_0^2)d\mathbf{r} = 4a_1|W(\sigma)|^2 = 0, \qquad |\vec{\lambda}_T| = 2\pi/|\mathbf{k}_T| \quad (12.50)$$

The nontrivial solution of this equation is $a_1 = 0$. From (12.49) this implies $\mathcal{Y}_1 = 0$. It is then easy to solve (12.48) and to obtain the solution

$$u_1 = -(2X_{T\pm}|W|^2/\delta^2) + [W_1e^{i\mathbf{k}_T\mathbf{r}} - (X_{T\pm}W^2/9\delta^2)e^{2i\mathbf{k}_T\mathbf{r}} + c.c.] \quad (12.51)$$

The function $W_1 \equiv W_1(\sigma)$ is the general solution of the homogeneous problem. It is not determined at this order of the perturbation expansion.

At the second order in η we obtain from the Swift–Hohenberg equation (12.43)

$$\mathcal{L}u_2 = X_{T\pm}(6a_2u_0 + 6u_0u_1 + 3u_0^2) + u_0^3 + \partial u_0/\partial\tau \qquad (12.52)$$

and from the steady state equation (12.44)

$$4\mathcal{Y}_2 = (3X_{T\pm}^2 - C)a_2 \tag{12.53}$$

Expressing that the right-hand side of (12.52) must be orthogonal to u_0 (solvability condition) yields an equation for the unknown function W

$$\partial W/\partial\sigma = -(8X_{T\pm}\mathcal{Y}_2/\delta^2)W + (87 + 38C/\delta^2)|W|^2 W/9 \tag{12.54}$$

In terms of the unscaled amplitude $\eta W = A\exp(i\phi)$, this last equation can be written as

$$\partial A/\partial t = (h_0 + h_1 A^2)A, \qquad \partial\phi/\partial t = 0 \tag{12.55}$$

$$h_0 = 8X_{T\pm}(\mathcal{Y}_{T\pm} - \mathcal{Y})/\delta^2, \qquad h_1 = (87 + 38C/\delta^2)/9 \tag{12.56}$$

This equation has two steady solutions

- The trivial solution $A = 0$ that corresponds to the homogeneous solution
- The nontrivial solution $A = \pm\sqrt{-h_0/h_1}$ that exists iff $h_0 h_1 < 0$

An exact solution of the amplitude equation (12.55) is easily obtained and indicates that the trivial solution is stable for $h_0 < 0$; that is, $\mathcal{Y} < \mathcal{Y}_{T-}$ or $\mathcal{Y} > \mathcal{Y}_{T+}$. The nontrivial solution is stable for $h_0 > 0$ and therefore $h_1 < 0$. This last equality implies

$$C < -87\delta^2/38 \tag{12.57}$$

The picture that emerges from this analysis is that in the monostable domain (because C must be negative), the Turing bifurcation is supercritical. The branch of inhomogeneous solutions that emerges from the Turing bifurcation is stable close to the bifurcation point where it exists for $\mathcal{Y}_{T-} < \mathcal{Y} < \mathcal{Y}_{T+}$.

To analyze the domain $C > -87\delta^2/38$, which includes the bistable homogeneous solution, we need to continue the perturbation expansion in powers of η two orders higher: The next approximation for the amplitude equation contains fifth-order terms. Using the polar decomposition of $\eta W = A\exp(i\phi)$ yields the amplitude equation

$$\partial A/\partial t = (g_0 + g_1 A^2 + g_2 A^4)A, \qquad \partial\phi/\partial t = 0 \tag{12.58}$$

$$g_0 = 8X_{T\pm}(\mathcal{Y} - \mathcal{Y}_{T\pm})/\delta^2$$

$$g_1 = (87 + 38C/\delta^2)9 + 16X_{T\pm}(\mathcal{Y} - \mathcal{Y}_{T\pm})(24\delta^2 - 16C)/(243\delta^2)$$

$$\qquad - 32(\mathcal{Y} - \mathcal{Y}_{T\pm})^2(1864\delta^4 + 4881\Delta^2 C - 958C^2)/(6568^{10})$$

$$g_2 = (2438^8 - 2825288^4 + 357008^2C + 43292C^2)/(155528^6)$$

This time, the steady state solutions of equation (12.58) are either the trivial solution or the roots of the biquadratic

$$A = 0, \qquad A_\pm^2 = \left(-g_1 \pm \sqrt{g_1^2 - 4g_0g_2}\right)/(2g_2) \qquad (12.59)$$

The trivial solution coincides with A_- at $\mathcal{Y} = \mathcal{Y}_{T\pm}$. In the half plane $(A > 0, \mathcal{Y})$, the nontrivial solutions represent a parabola with a limit point at $A_- = A_+$. The ordinate of the limit point is given by $g_1^2 = 4g_0g_2$, which is an implicit equation for $\mathcal{Y}_{L\pm}$.

A linear stability analysis of the steady solutions (12.59) indicates that the trivial solution (homogeneous solution) is stable for $\mathcal{Y} < \mathcal{Y}_{T-}$ and $\mathcal{Y} > \mathcal{Y}_{T+}$, and that A_- is always unstable. The stability of the upper branch A_+ has not been assessed analytically.

In the preceding analysis, we have made an assumption that has deep consequences on the nature of the solutions. Indeed, we have assumed that the solution (12.47) is characterized by only one wave vector. However, the general solution of (12.46) is

$$u_0 = \sum_{n=1}^{N} \left(W_n(\sigma)\exp(i\mathbf{k}_n\mathbf{r}) + c.c.\right), \qquad |\mathbf{k}_n| = |\mathbf{k}_T| \qquad (12.60)$$

If $N = 1$, the only patterns that can appear are stripes. They correspond to periodic oscillations in space. If $N = 2$, there are rhombi with $|\mathbf{k}_1 + \mathbf{k}_2| \neq |\mathbf{k}_T|$. If $N = 3$, the solutions are hexagons with $\mathbf{k}_1 + \mathbf{k}_2 + \mathbf{k}_3 = \mathbf{0}$. From the analytic point of view, the main result that has been obtained for $N = 3$ can be explained as follows [8]. There is a domain of input field amplitude where stripes and hexagons coexist. That is, when each of these patterns is subjected to a perturbation that has the symmetry of the pattern, it is stable. To resolve this ambiguity, a *relative stability analysis* has to be made, where the stability of a pattern is tested with respect to perturbations having the symmetry of the other pattern. In this case, it was shown that the bistability between stable stripes and stable hexagons disappears and leads to an instability point where hexagons lose their stability and stripes become stable.

12.5.3 Numerical results

No analytical study has been able yet to unveil the richness of the Swift–Hohenberg equation that appears when it is simulated numerically, though the number of analytical studies increases (see, for instance, [5] and [9]). Therefore, in this section, we break our rule to limit the discussion to analytic results and we present a variety of solutions of the Swift–Hohenberg equation (12.37) obtained numerically. In all cases, periodic boundaries are imposed on the transverse dimension(s).

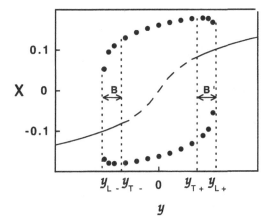

Figure 12.1 Bifurcation diagram for the Swift–Hohenberg equation (12.37) in one transverse dimension with $\delta = 0.1$ and $C = -0.001$. The solid (dashed) line represents the stable (unstable) homogeneous steady state solution. The dots above (below) the homogeneous solution are the maxima (minima) of the stable space-periodic solutions. Courtesy of M. Tlidi.

First, we display in Figure 12.1 a typical bifurcation diagram obtained for one transverse dimension with $\delta = 0.1$ and $C = -0.001$. Because $\delta = 0.1$, the critical value that separates the supercritical from the subcritical Turing bifurcation is $C_c = -87\delta^2/38 \cong -0.023$. Therefore the value $C = -0.001$ is in the monostable domain and the Turing bifurcation is subcritical. Consequently, there are two symmetric domains of bistability, labeled B and B' on the figure, in which there is coexistence of the stable homogeneous solution and an inhomogeneous solution. This inhomogeneous solution is a simple space-periodic function whose maxima and minima are represented by dots in the figure. This branch of space-periodic solutions is stable between the two limit points at $\mathcal{Y}_{L\pm}$.

In 2-D, there are more possibilities of spatial patterns. As an example, displayed in Figure 12.2 are three pure solutions and one mixed solution. Figure 12.2(a) represents hexagons with the maximum of the intensity in the center of the hexagons (the so-called $H0$ hexagons). Another type of hexagons, the $H\pi$, are displayed in Figure 12.2(c), where the center of the hexagons has a minimum of intensity and the maximum is along the hexagon boundaries. The origin of these two solutions is easy to understand. The hexagon amplitudes $W_j(j = 1, 2, 3)$ are decomposed in polar coordinates $W_j = |W_j| \exp(i\phi_j)$. The existence of a steady state imposes that $\sin(\phi_1 + \phi_2 + \phi_3) = 0$. The two possible solutions, $\phi_1 + \phi_2 + \phi_3 = 0$ or π, determine the two classes of hexagons. Figure

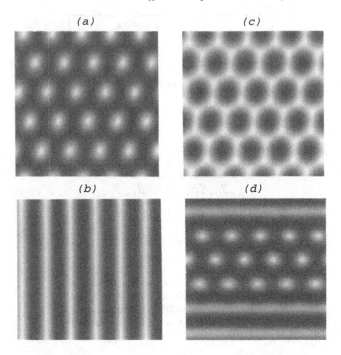

Figure 12.2 Sample of the space-periodic solutions of the Swift–Hohenberg equation (12.37) for $\delta = 0.1$ and $\mathcal{C} = -0.005$. (a) $H0$ hexagons $(\mathcal{Y} = -0.0004)$; (b) stripes $(\mathcal{Y} = -0.00001)$; (c) $H\pi$ hexagons $(\mathcal{Y} = 0.0004)$; (d) mixed pattern $(\mathcal{Y} = -0.0001)$. For the integration, the spatial grid is 60×60. Courtesy of M. Tlidi.

12.2(b) shows a stripe pattern characterized by a single wave vector. Finally, Figure 12.2(d) is an example of a mixed pattern that combines $H0$ hexagons and stripes.

In general, the periodic spatial solution that emerges from the Turing bifurcation does not form a continuous stable branch of solutions. Secondary instability points are generic, and a higher degree of complexity appears as one tries to follow the simple spatial patterns. Often, these secondary bifurcations induce defects in the simple patterns. This means a spontaneous symmetry breaking.

Examples are shown in Figure 12.3, where three types of defects are displayed. In Figure 12.3(a), a hexagonal pattern is modified by pairs of "penta-hepta" defects, where some spots are surrounded by either five or seven spots. In Figure 12.3(b), stripe patterns with different orientations coexist. In Figure 12.3(c), a double dislocation of a stripe pattern has occurred.

The previous example of defects was obtained for a square transverse section. However, the transverse geometry must influence in some way the pattern

(a)   (b)   (c)

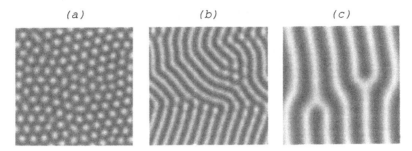

Figure 12.3 Defects of the space-periodic solutions of the Swift–Hohenberg equation (12.37) for $\delta = 0.1$ and $\mathcal{C} = 0.025$: (a) penta-hepta defects in an $H0$ pattern ($\mathcal{Y} = -0.0001$); (b) stripe domains with different orientations ($\mathcal{Y} = -0.00008$); (c) a pair of dislocations in a stripe pattern ($\mathcal{Y} = 0$). For the integration, the spatial grid is 100×100 for (a) and (b) and 60×60 for (c). Courtesy of M. Tlidi.

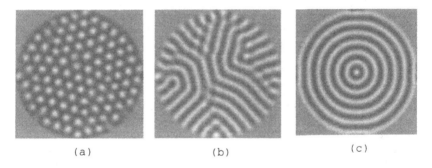

(a)   (b)   (c)

Figure 12.4 Defects of the space-periodic solutions of the Swift–Hohenberg equation (12.37) for $\delta = 0.1$ and $\mathcal{C} = 0.025$ and a circular transverse section. (a) penta-hepta defects in an $H0$ pattern ($\mathcal{Y} = -0.0001$); (b) multi-stripe domains ($\mathcal{Y} = -0.0008$); (c) target pattern ($\mathcal{Y} = -0.0008$). The spatial grid for the square is 100×100. The equation is integrated with the constraint that the area outside of the circle is the homogeneous steady state for all time. The periodic boundary is applied to the square. Courtesy of M. Tlidi.

selection. This is shown clearly in Figure 12.4 where a circular transverse section has been imposed.

Figure 12.4(a) displays pairs of penta-hepta defects in an $H0$ structure. Figure 12.4(b) displays multiple stripe domains. Note that the stripes end up orthogonally to the circular boundary. Figure 12.4(c) displays a "target" pattern that coexists with the multistripe pattern of Figure 12.4(b). If the Turing bifurcation is subcritical, there is a domain of bistability between the homogeneous and inhomogeneous solutions. In that domain, another

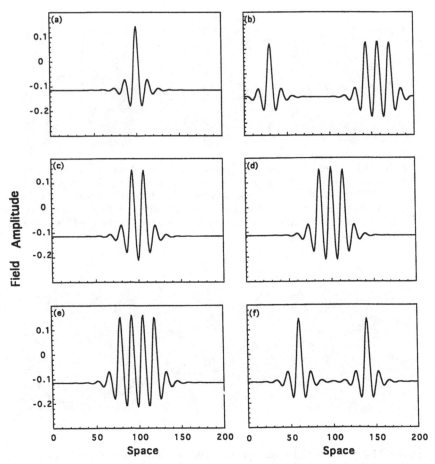

Figure 12.5 Localized solutions of the Swift–Hohenberg equation (12.37) in one transverse dimension for $\delta = 0.1, C = -0.001$, and $\mathcal{Y} = -0.0004$. All six figures are obtained with the same parameters but different initial conditions. Courtesy of M. Tlidi.

type of solution set may exist: the localized solutions. For the 1-D Swift–Hohenberg equation (12.37), this type of solution is displayed in Figure 12.5.

The localized solutions connect the homogeneous solution (flat part of the curves) and the periodic solution with which it coexists. The number of peaks present in a given localized solution is arbitrary (within the limits set by the boundary conditions). The six solutions displayed in Figure 12.5 differ only by their initial conditions; all other parameters are the same. Such a behavior suggests a multisoliton-type of solution, or more generally, homoclinic solutions. This was proved to be the case in one transverse dimension by Glebsky and Lerman [10]. An introduction to their proof will be given in the next section.

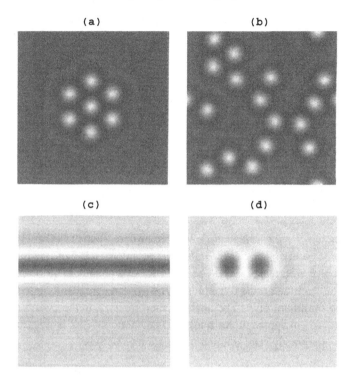

Figure 12.6 Localized solutions of the Swift–Hohenberg equation
(12.37) in two transverse dimensions for $\delta = 0.1$ and $\mathcal{C} = 0.025$: (a)
localized hexagon ($\mathcal{Y} = -0.0005$); (b) random distribution of localized spots obtained for the same parameter as in (a) but with a different initial condition; (c) metastable localized stripe ($\mathcal{Y} = 0.0001$) that
yields in the long time limit the pattern; (d) pair of localized spots. The
grid size is 100×100 for (a) and (b), 60×60 for (c) and (d). Courtesy
of M. Tlidi.

In two transverse dimensions, the existence of localized structures is more difficult to assess analytically. Numerically, they have been found, as shown in
Figure 12.6.

The coexistence of different localized structures is evident from Figures
12.6(a) and (b), which differ only by the initial condition. Figure 12.6(c) displays a localized stripe that is in fact metastable: In the very long time limit,
it disappears to produce the two spots shown in Figure 12.6(d), which form a
stable localized structure.

The interest of the localized structures stems from the coexistence of a
large number of them. Because substracting or adding an extremum (i.e., a
spot in the transverse profile) means replacing one pattern/solution by another

pattern/solution that has the same domain of stability, one may expect that they could be used to encode information. Adding or substracting a maximum then becomes a write or erase operation on the "memory."

12.6 The Glebsky–Lerman hamiltonian

A rather unique feature of the Swift–Hohenberg is that it is a gradient form; that is, it derives from a potential. Indeed we can write (12.36) as

$$\partial u / \partial \tau = -\delta F / \delta u$$

$$F[u] = -\int \left\{ \frac{p}{2} u^2 + \frac{q}{3} u^3 - \frac{1}{4} u^4 - \frac{1}{2}[(1 + \mathcal{L}_\perp)u]^2 \right\} d\mathbf{r} \quad (12.61)$$

The parameters p and q have been defined in the derivation of (12.36): $p = 1 + (\mathcal{C} - 3\mathcal{E}_0^2)/3\delta^2$ and $q = -\mathcal{E}_0 \sqrt{3}/\delta$. In these equations, $\delta F / \delta u$ is the functional derivation of F with respect to u. The existence of a potential $F[u]$ implies that if u is a stable attractor, the dynamics of the system consists of the relaxation toward the minimum of F. The analysis then focuses on the possible steady states. One way to approach this question is by noticing that, in one transverse dimension, the steady state solution of equation (12.36) is

$$u'''' + 2u'' - (p - 1)u - qu^2 + u^3 = 0 \quad (12.62)$$

with $u' \equiv \partial u / \partial x$. It was shown by Glebsky and Lerman [10] that this equation can be written as a Euler–Lagrange equation

$$\frac{d^2}{dx^2} \frac{\partial L}{\partial u''} - \frac{d}{dx} \frac{\partial L}{\partial u'} + \frac{\partial L}{\partial u} = 0 \quad (12.63)$$

for the lagrangian

$$L = \frac{1}{2}(u'')^2 - (u')^2 - \frac{1}{2}(p - 1)u^2 - \frac{q}{3}u^3 + \frac{1}{4}u^4 \quad (12.64)$$

From this result follows a hamiltonian formulation of the stationary solutions of the Swift–Hohenberg equation. We define two variables u and $v = u'$ together with two moments $p_v = u''$ and $p_u = -2u' - u'''$. These variables and momenta are connected via the Hamilton equations

$$u' = \partial H / \partial p_u, \qquad v' = \partial H / \partial p_v, \qquad p_u' = -\partial H / \partial u, \qquad p_v' = -\partial H / \partial v$$

$$(12.65)$$

with the hamiltonian H defined as

$$H = v p_u + v^2 + \frac{1}{2} p_v^2 + \frac{p-1}{2} u^2 + \frac{q}{3} u^3 - \frac{1}{4} u^4 \qquad (12.66)$$

The value of this formulation is that it enables the use of the powerful methods that have been set up to study hamiltonian dynamics. In particular, small amplitude solutions of the Swift–Hohenberg equation (12.62) require the analysis of the hamiltonian equations in the vicinity of the origin $(u, v, p_u, p_v) =$ (0, 0, 0, 0). This is precisely what has been done by Glebsky and Lerman, who proved that in the vicinity of $p = 0$; that is, $3\delta^2 + \mathcal{C} - 3\mathcal{E}_0^2 = 0$, a countable infinite set of homoclinic solutions exist. They are the multisoliton solutions of the steady-state Swift–Hohenberg equation (12.62), also known as *localized structures,* displayed in Figure 12.5.

References

[1] N. B. Abraham and W. J. Firth, eds., *Transverse effects in nonlinear optics,* *J. Opt. Soc. Am. B* **7** (1990) 948–1157 and 1264–1373.

[2] L. A. Lugiato, *Spatio-temporal structures I, Physics Reports A* **219** (1992) 293.

[3] C. O. Weiss, *Spatio-temporal structures II, Physics Reports A* **219** (1992) 311.

[4] L. A. Lugiato, *Nonlinear optical structures, patterns and chaos. Chaos, Solitons & Fractals* **4** (1994) 1251–1844.

[5] M. C. Cross and P. C. Hohenberg, *Rev. Mod. Phys.* **65** (1993) 851.

[6] P. Collet and J.-P. Eckmann, *Instabilities and Fronts in Extended Systems* (Princeton University Press, Princeton, 1990).

[7] M. Le Berre, E. Ressayre, and A. Tallet, *Quantum & Semiclass. Opt.* **7** (1995) 1.

[8] M. Tlidi, M. Georgiou, and P. Mandel, *Phys. Rev. A* **49** (1993) 4605.

[9] A. V. Gaponov-Grekhov and M. I. Rabinovich, *Nonlinearities in Action* (Springer, Heidelberg, 1992).

[10] L. Yu. Glebsky and L. M. Lerman, *Internat. J. Nonlin. Sci.* **5** (1995) 424.

[11] M. Tlidi and P. Mandel, *Chaos, Solitons & Fractals,* **4** (1994) 1475.

Index